THE END

The End of the CBC?

DAVID TARAS AND CHRISTOPHER WADDELL

UNIVERSITY OF TORONTO PRESS
Toronto Buffalo London

© University of Toronto Press 2020
Toronto Buffalo London
utorontopress.com
Printed in Canada

ISBN 978-1-4875-9353-7 (cloth) ISBN 978-1-4875-9354-4 (EPUB)
ISBN 978-1-4875-9352-0 (paper) ISBN 978-1-4875-9355-1 (PDF)

Library and Archives Canada Cataloguing in Publication

Title: The end of the CBC? / David Taras and Christopher Waddell.
Names: Taras, David, 1950– author. | Waddell, Christopher, 1952– author.
Description: Includes bibliographical references and index.
Identifiers: Canadiana 20190203218 | ISBN 9781487593537 (cloth) |
ISBN 9781487593520 (paper)
Subjects: LCSH: Canadian Broadcasting Corporation. |
LCSH: Public broadcasting – Canada. | LCSH: Social media – Canada.
Classification: LCC HE8689.9.C3 T37 2020 | DDC 384.540971–dc23

We welcome comments and suggestions regarding any aspect of our publications—please feel free to contact us at news@utorontopress.com or visit us at utorontopress.com.

Every effort has been made to contact copyright holders; in the event of an error or omission, please notify the publisher.

University of Toronto Press acknowledges the financial assistance to its publishing program of the Canada Council for the Arts and the Ontario Arts Council, an agency of the Government of Ontario.

Canada Council
for the Arts
Conseil des Arts
du Canada

ONTARIO ARTS COUNCIL
CONSEIL DES ARTS DE L'ONTARIO
an Ontario government agency
un organisme du gouvernement de l'Ontario

Funded by the Financé par le
Government gouvernement
of Canada du Canada

Canadä

MIX
Paper from
responsible sources
FSC® C016245

This book is dedicated to our grandchildren, Thaddeus Taras and Liam Waddell, in the hope that they will enjoy long and happy lives.

Contents

Preface

This book is about three overlapping crises: the crisis that has enveloped the CBC, the crisis of news, and the crisis of democracy. They are all the result, to some degree, of the vast changes that have overtaken and consumed the media world in the last 10 to 15 years. The emergence of platforms such as Google, Facebook, Twitter, and Netflix; the hypertargeting of individual users through data analytics; the development of narrow, online-identity communities; and the blast of an attention economy that makes it more and more difficult for any but the most powerful media organizations to be noticed have changed the media landscape in dramatic ways. The effects on the CBC and on other Canadian media organizations have been shattering. To put it bluntly, news and journalism are in a deep crisis, for reasons that we will explain in considerable detail in the book.

Our argument is that the CBC, Canada's public broadcaster, has reached a crossroads. Years of budgetary uncertainty, a lack of policy vision by governments and by the CBC itself, and the brutality of the attention economy have taken a toll. For the CBC, the choices are stark. The public broadcaster will either be reimagined and reinvented or die a slow death on the outskirts of the media world. We suggest a way of going forward that would transform much of news and, as a consequence, public affairs in Canada.

We could not have undertaken this journey without help from others. We are indebted to the scholars who have paved the way in studying public broadcasting in Canada and Canadian media. They are, among others, Bart Beaty, Patricia Cormack, James Cosgrove, Brooks DeCillia, Monica MacDonald, Patrick McCurdy, Mary Jane Miller, Frank Peers, Mark Raboy, Wade Rowland, Paul Rutherford, Florian Sauvageau, Richard Schultz, Rebecca Sullivan, and Gregory Taylor. In addition, we benefited from a host of memoirs and books by former CBC executives, producers, and journalists. A number of CBC executives and journalists, as well as former colleagues of Christopher Waddell's, from his time at the CBC, spoke to us off the record, and we benefited from the work of media analysts who provided key perspectives on audience and budget numbers.

We owe a special debt of gratitude to the two outside reviewers engaged by the University of Toronto Press for their valuable feedback and suggestions and to our colleagues who took the time to read the manuscript and provided important insights and encouragement. We would like to thank Sean Holman and Gregory Taylor, in particular.

David Taras would like to thank Dean Elizabeth Evans and department chair Brad Clark at Mount Royal University for their patience and support. Christopher Waddell extends similar thanks to Carleton University's dean of the Faculty of Public Affairs, Andre Plourde; director of its School of Journalism and Communication, Josh Greenberg; and associate director Susan Harada for their support and to his faculty colleagues at the School, with whom he has frequently debated many of the issues we explore in this book.

We are also grateful to those with whom we have worked at the University of Toronto Press. Michael Harrison was the one who first brought us on board, and working with him was an immense pleasure. After Michael retired, we worked with Mat Buntin and then Marilyn McCormack, who have been extremely supportive and highly professional. Stephanie Stone deserves praise for the fine job she did in copy editing the manuscript.

Finally, David Taras would like to thank his wife, Joan, for her love and encouragement and our growing family: Matthew and Victoria; Joel, Asher, Matthew, and Devra; and David for bringing so much joy and happiness into our life.

Christopher Waddell would like to thank his wife, Anne, for her love and support and our family: Matthew and Maria and Kerry. He also adds a special thanks to his mother, Evelyn, who wrote children's novels for more than four decades under the name Lyn Cook and appeared in a children's program, *A Doorway to Fairyland*, on CBC radio in the 1950s. She made it to the age of 100 and was thrilled to hear about the plans for this book but sadly did not see its publication.

THE END OF THE CBC?

Chapter 1

Introduction

When discussing the Canadian Broadcasting Corporation, pundits and commentators rarely mention the fact that the public broadcaster has a unique mission in the Canadian media landscape. In their content analysis of newspaper coverage of the CBC, Brooks DeCillia and Patrick McCurdy found what they describe as "the sound of silence."[1] The CBC is reported on, attacked, lambasted, praised, and scorned by journalists and commentators, who almost never bother to mention that, unlike other media organizations, the CBC has a public service mandate and is intended as a "public good."[2] Almost never discussed is the fact that the CBC's mandate is to reflect Canadian identity, address its audiences as citizens rather than consumers, and carry out tasks that the private networks can't or won't perform. Instead, the CBC is depicted as getting in the way of private broadcasters, as wasting taxpayers' money, and as an afterthought rather than a main event. In short, the public broadcaster seems to be losing the battle of public perceptions that it needs to win to survive.

The CBC is a public broadcaster in three senses. It is largely supported by public funds (through an annual parliamentary appropriation); its mandate is to reflect the public and national interest; and at a time when Canadians are buried by an avalanche of subscriptions, pay-per-view options, and micropayments, the CBC's main TV and radio services and websites are freely available. On this last point, it should be noted that cable channels such as CBC News Network and the French-language equivalent, Ici RDI (Réseau de l'information), are supported through subscriptions. The fact that the CBC is still largely freely available is especially important because with many, if not most, Canadians just one or two paycheques away from falling off a financial cliff, their media world is getting smaller rather than larger.

Founded in 1936, the CBC/Société Radio-Canada (Radio-Canada being its French-language arm) has been one of the great nation-building successes in Canadian history. Arguably, the founding of the CBC in English and French stands alongside Sir John A. Macdonald's National Policy of 1879, which included the construction of a railway that stretched from coast to coast; the creation of social programs such as old age pensions, employment insurance, and medicare in the post–World War II period; and the Canadian Charter of Rights and Freedoms in 1982 as a basic building block of the Canadian state and, indeed, of Canadian citizenship.

Public broadcasters have played this state-building role almost everywhere in the world. A German scholar, Christina Holtz-Bacha, contends that national and cultural integration and the need to provide "social cement" were the driving forces behind the creation of public service media throughout post-war Europe after 1945.[3] In Germany, in particular, where broadcasting is the responsibility of state governments (*Länder*), public broadcasters were to act as a check on the power of the federal government and as a watchdog, ensuring that a Nazi-style dictatorship would never emerge again. Scandinavian scholars Trine Syvertsen, Gunn Enli, Ole Mjos, and Hallvard Moe believe that public broadcasters in Nordic countries have provided the "social glue" that has allowed the goals of the welfare state, including equality, solidarity, and belonging, to be largely achieved.[4] The goal of public broadcasters in what they call the "media welfare state" is to set the standard for private broadcasters by raising the quality of programming and audience expectations.

While public broadcasters share roughly the same social goals almost everywhere, they differ depending on history and circumstances. For example, in Ireland, RTÉ acts as a protective shield against the headlong assault of British broadcasting and one of the only means for the Irish to hear their own voices. In Australia, where much of the media is owned by ardent conservative Rupert Murdoch, the Australian Broadcasting Corporation not only spans a large country but also provides citizens with access to a diversity of viewpoints. While never explicitly discussed in documents or policy statements, providing a counterweight to Murdoch has become critical to Australian democracy.

The CBC's multifaceted and ambitious mandate is itemized in the Broadcasting Act of 1991. The public broadcaster is expected to be "predominantly and distinctly" Canadian, reflect the country and its regions, contribute to "national consciousness and identity," broadcast in each official language, stimulate "the flow and exchange of cultural expression," and reflect Canada's multicultural and multiracial character. In other words, the CBC's mandate is to be all things to all people all the time. We will argue that, without the necessary resources, and facing fierce competition on all fronts, the mission given to the CBC by the Broadcasting Act seems more like a distant dream, an illusion, than an achievable reality.

The CBC was to be a connecting link and a unifying voice in a country challenged by vast distances, sharp and often painful linguistic and regional divides, and the pervasive influence of American culture. The distances alone are breathtaking. Canada stretches across a quarter of the world's time zones, borders three oceans, and is so vast that the United Kingdom and Ireland can fit comfortably into Saskatchewan, a single province, while all of Germany can fit neatly into Alberta. These distances are so dizzying that they are beyond the comprehension of people in most other countries. Moreover, most of Canada's population resides in an archipelago of city states that stretches for thousands of kilometres just north of the US border: Halifax, Montreal-Ottawa, Toronto-Hamilton, Winnipeg, the Calgary-Edmonton corridor, and Vancouver-Victoria.

The sensibilities of Montreal, however, are very different from those in St. John's, Calgary, or Vancouver. While the great national unity crises seem to have passed, the country's unity can never be taken for granted. Canada is a successful nation in many ways, but it has also faced great dangers. The 1995 referendum on Quebec sovereignty came close to breaking Canada apart; at the very least, the country suffered a national nervous breakdown. Moreover, reaching agreement among the provinces and territories on even minor issues is often mind-bogglingly difficult, if not impossible, and relations with Indigenous communities have often been broken and tarnished by misunderstanding and injustice.

To make matters even more complex, Pierre Trudeau's famous observation that sharing a continent with the Americans was the equivalent of being in bed with an elephant is especially true with respect to culture and entertainment. Hollywood dominates the Canadian cultural scene to such a degree that John Meisel, a former chair of the Canadian Radio-television and Telecommunications Commission (CRTC), once observed that Canadian culture is the minority culture in Canada. The majority culture is American. Indeed, he went so far as to claim that "inside every Canadian there is in fact an American—the degree of that Americanness varying from person to person depending on their values, education and exposure to popular culture."[5] Whether one agrees with Meisel or not, American influence is arguably greater in Canada than in any other developed country.

Amid these challenges, the CBC attempts to create a cultural experience that ties the country together. As philosopher John Ralston Saul observed some time ago, "Everybody who is smart in bureaucracies and governments around the Western world now knows that public broadcasting is one of the most important remaining levers that a nation state has to communicate with itself."[6] Public broadcasting is about creating an "imagined community," to use Benedict Anderson's famous phrase, where people come to identify with people and events that are beyond their personal experiences.[7] Listeners and viewers who may never have been to Ottawa, travelled to the distant North, or visited Vancouver Island can experience and gain an emotional connection to the country in which they live and learn about the challenges that it faces. Broadcasting is not just a toaster that emits sounds and pictures, to use former chair of the US Federal Communications Commission Mark Fowler's famous description. It is nothing less than the spinal cord that links history, meaning, and identity. Today, supporters of public broadcasting in Canada and elsewhere also believe that its goal should be to build "social capital" by educating and uplifting citizens and ensuring that they have the essential information they need to make decisions about their lives. Public broadcasting is all about "bridging," "witnessing," and "connecting."

In practical terms, there is the need to broadcast across languages, distances, and communities. The CBC broadcasts in English, French, and eight Indigenous languages and must have a national, regional, and local presence. In 2018, it operated 27 conventional, over-the-air TV stations, 88 radio stations, and a large flotilla of websites. Moreover, since the 1970s, Canada has absorbed more immigrants per capita from more places than any other country in history. While this multicultural mosaic has become so much a part of our daily lives, particularly in urban Canada, that we no longer notice it or think that it is unusual, the reality is that many Canadians are still discovering their new country. For them, the Canadian media is not only a transmission line but also a port of entry. Scholars at Université Laval found that, unlike the experience of previous immigrants, who were largely cut off from the societies that they left behind, today's globalized media allows new Canadians to maintain contact with their former countries so that multiple allegiances and identities are easier to maintain.[8]

The heart of the CBC's mandate is that it's expected to present the Canadian view of the world in news, entertainment, and drama. While, at first blush, being relentlessly Canadian might sound easy to accomplish, the obstacles that lie in the path of Canadian productions and in finding audiences for those productions are multiplying. When it comes to drama, for instance, Canadian programming is more expensive to produce and far less lucrative than broadcasting the glossy and star-studded American programs that are the advertising-revenue bread and butter of Canada's private broadcasters. American shows dominate their prime-time schedules, and they can be bought off the shelf in Hollywood for far less than the cost of producing an equivalent Canadian show. Many of them are now available to Canadians through streaming services such as Amazon Prime, Netflix, Disney Plus, and Apple TV+. Canadian production is but a drop in an increasingly large programming sea.

Herbert Simon, the Nobel Prize–winning economist, did much of the pioneering work on attention economics. As he famously observed, "What information consumes is rather obvious: it consumes the attention of its recipients. Hence a wealth of information creates a poverty of attention...."[9] Today, scholars such as Matthew Hindman, James Webster, and Timothy Wu, among a host of others, have made the attention economy a mainstay of their research, with each arguing that the battle for attention is the key to understanding how the Internet works.[10] The vast infinity of choices available to users means that media platforms cannot survive unless they learn how to use the technologies of persuasion to attract audiences and prevent them from migrating elsewhere. Arguably, all news, politics, and commerce have been changed by the fact that the game of attracting and retaining audiences, which once came automatically to almost all media organizations, now has to be fought for with little guarantee that the losers will emerge with anything but the scraps.

We will argue that the new attention economy is a ruthless and merciless world ruled by instant auctions and bidding wars, one in which mammoth corporations such as Netflix, Amazon Prime, Disney Plus, and AT&T are able to dominate schedules with blockbuster movies, sports events, and TV series. In some ways, it has never been more difficult to attract the attention of Canadians, even for the best Canadian shows.

This is despite the elaborate system of defensive walls erected by the federal government to protect Canada's cultural sovereignty. These include the CBC; the National Film Board of Canada; the Canada Media Fund, which subsidizes and invests in Canadian film and TV productions; Canadian content legislation; tax laws that discourage Canadian companies from advertising in American media; the Canada Periodical Fund; and a myriad of other subsidies and tax credits.

In addition to this elaborate wall of defences and subsidies, simultaneous substitution, a policy that allows cable and satellite providers to block American advertising and replace it with ads sold by Canadian broadcasters when the same American shows are being aired concurrently on Canadian and American networks—in effect, creating a right-of-way for Canadian advertisers—keeps hundreds of millions of dollars in the country that would otherwise have left. Provincial governments also subsidize film and TV production, largely in an attempt to attract jobs to their provinces.

This means that the very notions of private and public broadcasting are misleading. The CBC, for instance, has traditionally relied on advertising for roughly a quarter of its budget, depends on independent producers for much of its programming, and as is the case with *Hockey Night in Canada*, is a flag of convenience for Rogers, a private broadcaster that controls that program's content and receives all its advertising revenue. While the CBC is a public broadcaster, it has many private elements.

Canada's private broadcasters are, in fact, private-public hybrids, with all the advantages of public support but without the obligations of a public broadcaster. Private broadcasters benefit from generous subsidies, albeit indirectly, through the independent producers whose shows they buy. And they are virtually guaranteed to have their licence renewed by the CRTC, which regulates the communications highway in Canada, regardless of how dismal or inadequate the broadcasters have been in meeting their programming obligations. In fact, the bar has arguably been set so low that one wonders what it would actually take for a station or channel to lose its licence. Arguably, few private broadcasters could survive without heavy doses of public funding and government subsidies.

It's important to note that the CBC is not the only public broadcaster in Canada. TV Ontario, Télé-Québec, and the Knowledge Network in British Columbia have similar mandates at the provincial level; and one can argue that APTN, the Aboriginal Peoples Television Network, which is a private corporation, and CPAC, the Cable Public Affairs Channel, which is owned by Canada's cable providers and operates as a not-for-profit broadcaster, exist in a kind of twilight zone between private ownership and public service.

Our main argument in this book is that the CBC has come to the end of its rope in terms of carrying out the tasks of a public broadcaster. Decades of funding cuts, political interference, decisions to boost private-sector broadcasters and downgrade the CBC, and poor management have taken a heavy toll. We will show in the next section of this introduction that the CBC exists in a kind of limbo, maintaining a presence everywhere but increasingly unable to compete anywhere. The main problem, however, is that the CBC was established during a time of media scarcity, when it could play a vital role in creating media experiences that didn't exist in most places in Canada. Unfortunately for the CBC, the ground has shifted dramatically; we are now in a time of hyper-media abundance. In the new attention economy, there are few guarantees that anyone will be watching or listening to anything for very long.[11]

Even Herbert Simon, whose work in the 1970s is fundamental to attention economics, would be astonished by the avalanche of information that people now have at their disposal—not only every day but also every minute of every day.[12] The great dilemma facing the public broadcaster is that as the Canadian public becomes more fragmented, elusive, globalized, and ever shifting, the CBC is finding it increasingly difficult to reach the very audiences that its existence depends on.

We argue that, in the midst of the revolutionary changes in the media world that have occurred in the last 10 to 15 years—what David Taras has previously described as "media shock"—the public broadcaster has had increasing difficulty maintaining a foothold.[13] Like all broadcasters and media organizations, the CBC is struggling to adapt to a world shaped by Google searches, Facebook posts, Twitter bursts, data mining, Internet streaming, immersive videos, memes, mashups, pirating, ad blocking, and the ubiquity of mobile devices—with few of the resources needed to survive in this turbulent sea.

The crisis of Canadian broadcasting is complicated by the fact that the digital revolution has also created new media delivery systems. First, media goliaths such as YouTube, Facebook, Google, Netflix, and Disney are at once content providers as well as hosts for other media. While the CBC can reach its audience directly through regular broadcasting, it has to navigate through these new gateways as well. Unlike the old broadcasting system, these delivery systems exist outside Canadian control and regulation.

Second, we are now in a world of "spreadable media," or "one-to-many" communication, where every user has the capacity to be their own broadcaster. As Henry Jenkins and his colleagues have put it, "If it doesn't spread, it's dead."[14] While the flow of ideas and images still remains largely top down, much now depends on "grassroots intermediaries"—who bounce, redact, comment on, tweet about, meme, like, disrupt, and share what they see and listen to with networks of friends and people who share their interests. The user has, in some ways, become the producer. Spreadable media very much depends on using platforms such as YouTube, Facebook, WhatsApp, Snapchat, and Instagram, and it's increasingly difficult for CBC stories to be noticed amid the endless tidal waves of images, apps, posts, programs, videos, texts, and messages that crash across the Internet every single minute. In the end, what spreads depends on the populist instincts of the public and on decisions made by tens of millions, if not hundreds of millions, of individual users. As CBC shows and stories become less popular, or if they lose their edge, they become less spreadable.

Another important shift is that during the period of media scarcity that existed when the CBC was established, private broadcasters played a secondary role. For many decades, the CBC was the engine that carried and drove the system. Today, the balance of power has shifted. In a kind of jujitsu flip, the privates have gained the upper hand. While the CBC still plays a vital role, it carries out tasks that the private broadcasters find too expensive or difficult to fulfill. The public broadcaster has become a kind of default button, carrying programming that, by its very nature, is unlikely to attract a mass audience. The CBC is the fallback network, the supplementary broadcaster in Canada. Although it is still expected to cover national events that commercial media might otherwise not report on in any real depth, showcase Canadian talent and achievements, and tell Canadian stories when telling those stories is often expensive and difficult to produce, it has had surprisingly little political support in its battle with the private networks.

However noble and important the CBC may be as a signature of Canadian identity and belonging, we believe that unless it is reimagined and, in fact, reinvented, it will fade into a kind of zombie-like half-life. Much of its influence has already faded. In the next section, we will carry out a damage assessment by taking a tour through the CBC's many worlds, highlighting both its successes and the growing list of its losses and retreats. Our main conclusion is that the public broadcaster is caught in a vortex between two inescapable realities: relatively small audiences on the one hand and, on the other, a status quo that can no longer be sustained in the new attention economy. We are not suggesting that the CBC disappear; quite the opposite. We are suggesting that it needs to change its priorities, drop programming in some areas, and focus on others. Simply put, it needs to change its stripes if it is to survive in the new media landscape.

First, a disclaimer. This book is not about Radio-Canada, the CBC's French-language equivalent. While our comments about CBC as an overall corporation relate to its activities in all languages, in the following chapters references to the CBC mean its English-language services. While we refer to Radio-Canada in passing, we know little about the broadcaster and know little about the cultural sea in which it has to swim. While Radio-Canada has experienced some of the same shocks that have revolutionized English-language broadcasting, it is shielded, at least to some degree, from the full force of those blows by language and by the place that it holds in Quebec's political and cultural conversation. Our analysis deals almost exclusively with the English-language CBC and so do our proposals for changing the shape and scope of the public broadcaster. In the end, the country may end up with a public broadcaster that plays very different roles in English-speaking Canada than it does in Quebec; but this is already the case.

Skating on Thin Ice

Despite the noble vision of what public broadcasting can and should be, the CBC has become a shadow of its former self. It seems to have become a kind of Moore's Law in reverse. Intel co-founder Gordon Moore's famous dictum was that computing power would double about every two years—which, in fact, has been the case since the mid-1950s. The CBC, on the other hand, has had its English-language TV audiences cut by half every two decades. In 1967, it was so dominant that it had close to 49 per cent of the total national audience in English and French.[15] In 1970, for instance, the CBC TV station in Toronto captured 24 per cent of the TV-tuning audience as against 42 per cent for American stations and 22 per cent for CTV. This meant that shows such as *The Beachcombers*, *Front Page Challenge*, *Wojeck*, *Quentin Durgens, M.P.*, *The Wayne and Shuster Comedy Hour*, and *The White Oaks of Jalna* garnered huge audiences. Being able to attract sizable audiences also meant that the CBC was the fulcrum of a national star system, with actors, performers, and fictional TV characters enjoying wide followings across the country. It's not too much to say that, during this so-called golden age, the CBC was, in effect, the national home page and helped set the country's cultural and political agenda. To some degree, if it didn't happen on the CBC, it didn't happen at all.

By 1990, amid the maelstrom and hyper-fragmentation brought about by the explosion of cable and satellite channels, the CBC's share of the Toronto TV audience had plummeted by roughly half, to 13 per cent. In 2016–17, the endless kaleidoscope of YouTube videos, specialty services, apps, and streaming platforms had reduced the audience share of the CBC's main channel in prime time to 5.5 per cent, a figure boosted to some degree by large viewing numbers for the Summer Olympics in

Rio.[16] If one adds in the additional 1.6 per cent of the audience garnered by CBC News Network, the CBC's total television audience would be more than 7 per cent. More telling, perhaps, is the fact that the average Canadian spends only around 70 hours per year watching CBC TV, and about half of that viewing is hockey or foreign programming such as *Coronation Street* from the British Broadcasting Corporation (BBC).[17] Thirty-five hours a year is not a lot of viewing time given that the average Canadian watched a hefty 27 hours per week of conventional TV in 2016–17 and an additional 3.4 hours of Internet TV.[18] In some categories, the CBC's audience share is just 2 or 3 per cent or barely registers at all.

Arguably, the English-language CBC cannot lose any more of its TV audience without sinking into a kind of oblivion. It is running out of steps in its race to the bottom.

The numbers for Ici Radio-Canada Télé (television) are substantially better, but the trend line is roughly the same as for the English-language CBC. For instance, between 1976 and 1982, Radio-Canada commanded an average of 44 per cent of total French-speaking viewership.[19] With the advent of cable and Internet broadcasters, those numbers fell precipitously. But there is almost a day-and-night difference between how the English-language CBC and French-language network are doing in terms of audience share. In 2016–17, Ici Radio-Canada Télé scored a 20.9 per cent share in prime time.[20]

As we will point out in this volume, the CBC is caught in a deadly, three-pronged pincer movement. First, it finds itself in a content crisis, where airing and finding competitive Canadian content has become much more difficult and expensive. Second, the amount of competition has intensified on virtually every front. At the same time, audiences are much more splintered and elusive than could ever have been imagined even a short time ago. It's not clear that the CBC can find a way to escape this trap.

Despite the amount of time that Canadians spend on the Internet, TV remains a dominant medium and is essential to the CBC's very existence. Simply put, the CBC cannot lose the battle for TV and continue to exist. The facts are painfully obvious. According to the CRTC's 2018 annual *Communications Monitoring Report*, the average Canadian spends a staggering amount of time—over 30 hours a week—watching TV, whether on conventional channels or through streaming services.[21] This means that the average Canadian watches TV for at least a full day out of every week, or for close to 60 full days per year. If they live to 80, the average Canadian will have watched TV for 13 years out of their lifetime. While viewing hours per week are slipping among those who are between the ages of 18 and 34, dropping from 21.9 hours in 2010–11 to 16.5 in 2016–17, even for this younger cohort TV remains central, ever present, and inescapable.[22]

While Internet TV—or what is known as over-the-top viewing on platforms such as Netflix, YouTube Premium, Amazon Prime Video, Disney Plus, Bell's Crave, and HBO Now, among a small army of other providers—is growing, particularly among younger Canadians, at the time of writing it remains an adjunct to more traditional broadcasting. Viewers use Netflix and other platforms as an additional resource rather than as a substitute for the shows that they like to watch on the main channels or on cable. In 2017, for instance, close to 90 per cent of TV viewing time by Canadians aged 18 and above was on conventional and cable TV; only 11 per cent of viewing was on streaming services. Nonetheless, some 45 per cent of Canadians under 30 were "cord-cutters" or "cord-nevers"—meaning that they were no longer attached to traditional or cable TV. The figure for those aged 55 to 64 was 8.6 per cent. If there was ever a flickering warning light about what the future of TV is likely to look like, this is it.

There remain moments when the public broadcaster still exerts considerable influence and can assemble large audiences. This is especially the case when there is a crisis or a sensational news story or when shows have entered the cultural bloodstream in some way, but overall, the CBC's slide to the bottom of the attention economy is unmistakable. The stark reality is that most Canadians probably couldn't pick the stars of shows such as *Republic of Doyle, Schitt's Creek* (apologies to Eugene Levy and Catherine O'Hara), *Shoot the Messenger, Kim's Convenience, Baroness von Sketch Show,* or *Workin' Moms* out of a lineup. CBC journalists working on local supper-hour TV shows would probably not be recognized by people with whom they shared an elevator ride or who sat across from them at a coffee shop. In the new attention economy, or to put it differently, in the new economy of "continuous partial attention," to quote Linda Stone's much-used phrase,[23] the CBC's ability to get on people's screens in any meaningful way is fading. Although the situation is still markedly different in Quebec, as will be discussed in a few paragraphs, the same forces that are at work on the English-speaking side are also at work there—albeit at a slower burn rate.

The CBC's greatest problem, perhaps, is that, largely because of the sweeping mandate given to it by the 1991 Broadcasting Act, its programming appears on all platforms and in all genres. Forced to compete everywhere, the public broadcaster finds it increasingly difficult to compete anywhere. In programming area after programming area, genre after genre, it has lost ground and/or been displaced by others. One example is TV news. Audiences for *The National,* the public broadcaster's flagship English-language news show, have hemorrhaged since its glory days in the 1980s and 1990s. A rough estimate is that *The National* has lost half its audience since 1990, dropping more than a million viewers. Arguably, in almost any other environment, losses on this scale would have brought wholesale changes in management and on-air talent. Strangely, the public broadcaster clung to anchor Peter Mansbridge for 29 years, until the Mansbridge era finally sputtered to an end in 2017. A new format, anchored by a flotilla of four hosts—Adrienne Arsenault, Rosemary Barton, Andrew Chang, and Ian Hanomansing—which debuted in November 2017, is doing even worse in the ratings than the previous format did in the last year of the Mansbridge regime. The point is that while, for millions of Canadians, *The National* was once part of the "nightly rituals of citizenship," to borrow James Webster's phrase, and had extraordinary influence over the country's political and journalistic agenda, its voice and influence has shrunk substantially.[24]

The entire point of turning to a public broadcaster for news is to find insights, expertise, and in-depth analysis that are difficult to find anywhere else. Unfortunately, cutbacks have meant the downsizing or elimination of news bureaus, fewer investigative stories, and the elimination of a number of specialized beat reporters in areas such as the economy, environment, and justice who have the knowledge and contacts needed to dig more deeply into issues. Over time, the reasons to watch *The National* become fewer and fewer. While there are moments of true excellence, quality is often sacrificed in the name of what's trending today on social media, and the show descends far too often into the kind of "victim" news—the latest episodes of human suffering—that, however righteous it may make journalists feel about their role and however necessary it may be to tell some of these stories, that when repeated too often, leave audiences desperate to turn to almost anything else.

More important, perhaps, in the new attention economy, people receive news updates throughout the day and often feel little need for further updates at night. In many cases, they have already seen headlines, watched videos, shared stories, and read instant analysis. These demands make the CBC's job all that much more difficult than it was 10 or 15 years ago. In a society where people check their cellphones an average of three times an hour and spend between one and four hours a day on their phones, messaging, being prompted, reading articles, and playing games, the much-vaunted 24-hour news cycle has been shortened and pre-empted.[25] In this new context, the nature of journalism and audiences are changing dramatically.

In 2018, CBC News Network cancelled its daily business show, *On the Money*, reportedly because of a lack of funds. On a more fundamental level, audiences seemed unwilling to turn to the CBC when BNN Bloomberg, CNBC, the *Financial Post*, and countless other outlets inhabited the same space with substantially greater resources, connections to the corporate world, and expertise.

To make matters worse on the news front, the CBC long ago retreated from local television news, where it now offers only skeleton news shows that can barely scrape together a respectable audience. In some cities, audiences barely register. Local TV news is singularly important because it is a vital tie-in to local communities and a jumping-off point for evening viewing. People who watch the supper-hour news often stay on the same channel, and this behaviour affects viewing numbers for the entire evening. The sidelining of local news by the CBC over more than the past two decades has created an opening for private broadcasters, who have displaced it as the place to turn to when there is a local event or an important news story. Every few years, the CBC seems to come up with a new plan to reinvest in local coverage. Despite heroic efforts, it never seems to have lift-off.

Drama is another area where, at least with regard to English-language productions, the CBC has had difficulty competing. While it no longer produces its own dramatic shows, having outsourced that to independent production companies since the mid-1980s, it has had less than a stellar record of buying hit shows. While some CBC dramas have solid, respectable audiences, the competition in drama, especially with the arrival of HBO, Showtime, Netflix, and Amazon Prime, among a panoply of other cable and streaming services, is simply overwhelming. The harsh reality is that, in the age of peak TV, the CBC, despite all its efforts, hasn't had a show that has pierced the skin of popular culture for at least a generation. This is especially evident among younger Canadians, most of whom are unlikely to have watched or have even heard about these shows.

More worrying is that CBC English-language managers seem to have been strangely and inexplicably out of touch with what large segments of the audience want. For instance, the CBC has no equivalent of talk shows such as *Oprah*, *The View*, or *The Ellen DeGeneres Show* and has few representatives in reality TV, a genre that has exploded in popularity in the United States and the United Kingdom. Reality TV occupies much of the TV landscape and, most crucially, has become a focal point for audience online engagement and participation. CBC's failure on the talk-show front is especially mysterious given the success of *Tout le monde en parle* on Radio-Canada. When it comes to reality TV, the CBC experimented with shows such as *Battle of the Blades* (and is bringing it back after ending it years ago), *Making the Cut*, and *The Great Canadian Baking Show*, but thankfully never indulged in the ghost-sighting, body-make-over, real-life-police-chase, and emergency-room shows that drive so much of American cable TV. Nonetheless, there are reality shows that have captured the public imagination and raised awareness of issues, talents, and lifestyles. The problem may have been that network executives viewed reality TV shows as too trivial, trashy, and garish for public broadcasting. More or less dismissing an entire genre was arguably both a grievous error and a failure of the imagination.

The same sad tale has repeated itself in comedy. With the exception, perhaps, of the now departed *Rick Mercer Report*, the CBC hasn't had a runaway comedy hit for at least a decade. Fabled shows such as *Wayne and Shuster*, *The Kids in the Hall*, *The Newsroom*, and *Royal Canadian Air Farce*, which specialized in parody, sensational pranks, off-the-wall skits, and self-referenced Canadian culture and politics, were once the heart of the network schedule. Indeed, one night a week would be devoted just to comedy shows. *Twitch City*, which aired in the late 1990s, achieved a kind of cult following but never attracted a large audience. In separate studies, Patricia Cormack and James Cosgrave, and later Beverly Rasporich, have argued that these comedy shows reinforced Canadian values and gave legitimacy to the institutions that they mocked.[26] Today, most Canadians turn to other networks or go online if they want a good laugh.

Another major retreat, perhaps evacuation is a better term, is in sports programming. While the CBC still airs *Hockey Night in Canada*, Rogers holds the broadcasting rights to National Hockey League (NHL) games and controls every aspect of the show, including personnel choices and advertising revenue. Media analyst Ken Goldstein describes the broadcast as "Rogers on CBC," which is a painful but apt description.[27] The CBC's deal with Rogers allows it to maintain the fiction that it is still part of the game. The Canadian Football League (CFL), curling, soccer—including the FIFA World Cup—and baseball left the CBC long ago, and basketball, with its multi-ethnic and multi-racial, urban, hip following, never seemed to make it onto the CBC's screen. Spending on sports programming plummeted from close to $100 million in 2007 to a little more than $30 million in 2017, a small sliver of the more than $1 billion now spent by Canadian broadcasters on sports.[28]

CBC Sports seems to have become a kind of destination of last resort, where sports such as Stampede rodeo, IAAF Diamond League track and field, FIVB Volleyball, and Canadian Premier League soccer, which cannot find a spot on the main specialty sports channels, can find a home. We will discuss sports at much greater length later in the book.

The CBC's greatest loss, however, seems to be among younger viewers and listeners. To millennials, the CBC is increasingly an unknown land. They may visit on occasion, but they have little attachment to the network, although this situation differs somewhat in Quebec. In almost every category of watching and listening, the CBC's English-language audiences are older and increasingly geriatric. It is now the preferred broadcaster for 60- and 70-year-olds. Of all the warning signs, this one is the most significant.

The brutal reality is that younger media users inhabit a much different world than the one occupied by their parents. Traditional media in all its forms—books, magazines, conventional TV networks and TV news, newspapers, movie theatres, cable, and radio—is simply no longer as popular. Among many young people, this world has become non-existent, a ghost from another time, and the CBC is just another ghost.

Some Remarkable Successes

While there is enough bad news about the CBC to send media managers scurrying for the exits, there have also been remarkable successes. First, CBC Radio has been able to retain, and indeed expand, its audiences. According to the CRTC's *Communications Monitoring Report* for 2018, in English-speaking markets, CBC Radio One captured a 13 per cent audience share in fall 2017.[29] Ici Radio-Canada Première, its French-language equivalent, enjoyed a 17 per cent audience share during the same time period.[30] CBC Music and Ici Musique, which specialize in classical music on FM, attract only smallish audiences. But if their listeners are added to those of the main CBC channels, they boost audience numbers to 16 per cent on the English side and 22 per cent on the French side.[31] In other words, the CBC enjoys the kind of popularity in radio today that it had in television in the 1970s.

What's most remarkable, perhaps, is the loyalty of the CBC's audience. For many listeners, CBC Radio is the media space where they feel most at home, most completely Canadian. In particular, the CBC's morning and afternoon local drive-time shows are vital links in cities and communities across the country. *The Current* and *As It Happens* also deserve special mention as they play an important role in hosting national conversations on key issues facing the country. Strangely, CBC management has regularly cut budgets for radio, initially to buttress television and, more recently, to fund the corporation's digital strategy. It has eliminated jobs, reduced local noon-hour programming, and all but eviscerated radio drama. Radio is also subject to the same competitive pressures that we see in other traditional media. As we will discuss in a later chapter, streaming has brought about not only a cornucopia of choices but also hyper-targeting as well as customization and recommender systems, which are difficult for conventional radio broadcasters to compete against. If radio is the CBC's great pride and joy, it's likely to be less so in the future.

It's also the case that audiences for Ici Radio-Canada Télé remain relatively high. Its news and current events programs remain part of the journalistic mix in Quebec, and its journalists often break important stories on programs such as *Enquête*, its flagship investigative news show. It has spawned its own star system and has hit the ratings jackpot with *Tout le monde en parle*, a talk show that has become one of the great meeting places in Quebec society; high-voltage police dramas such as *Unité 9* and *District 31*; and zany comedies such as *Les pêcheurs*. The SRC remains part of the fabric of Quebec popular culture, and as such, it has an audience waiting for much of what it produces. Ici Radio-Canada Télé's saving grace is that, unlike the English-language CBC, it is shielded by language and a tight-knit Quebec identity from the full blast of American media competition.

The SRC also receives substantially more money per capita than its English-language counterparts. We are tempted to argue that there are two worlds of public broadcasting: one that has the resources needed to compete and one that can barely keep the lights on. It is also the case that mainly because of the larger investments that it makes in programming and promotion, Radio-Canada generates some 40 per cent of overall CBC advertising revenues.[32]

Richard Stursberg, a former head of the CBC's English-languages services, believes that taking money out of the rest of Canada to support Quebec programming has been a profound mistake. He believes that while English-Canadian media has had to face the American entertainment hurricane head on, "Québec culture has already won" the battle for survival. Nor are resources shared between the two networks. While Stursberg notes that programming does not translate easily because of differing visual styles and cultural cues, he was disturbed by the lack of interest that each side of the linguistic divide within the CBC/Radio-Canada has had for what the other side was doing. As he observed with respect to attitudes at CBC Toronto, "To the English side of the CBC, the French side is all but invisible."[33] The same seems to be true at Radio-Canada.

One key crossover, however, is in the stationing of journalists across the country. With one or two exceptions, which vary from time to time, the CBC is often the only news organization that has French-language reporters stationed in English Canada and English-language reporters stationed in Quebec outside Montreal. Admittedly, this is a small bridge, but it is one that would likely not exist without the CBC.

Another area of success is CBC North. Broadcasting to Yukon, the Northwest Territories, and Nunavut as well as to Cree speakers in Northern Quebec, CBC North offers programming in English and French as well as in eight Indigenous languages. Most programming consists of one-hour radio shows in each Indigenous language as well as TV news shows in English and French. While populations are sparse and the size of audiences is often difficult to gauge in an era of expensive or non-existent high-speed connections in isolated communities, the CBC remains a vital lifeline to communities across the great expanse of the North. For instance, the Northwest Territories covers 1,143,793 square kilometres, with a population density of 0.04 people per square kilometre. Infrastructure of any kind, including broadcasting, depends on government involvement. In its 2015 report on the CBC, the Senate Standing Committee on Transport and Communications noted that weather reports sometimes made a life-or-death difference to people in these isolated communities.

CBC TV's *The Fifth Estate*, a show dedicated to investigative journalism that has been on the air since 1975, has been a singular triumph. While it has almost never been a ratings success, its investigations into the death by suicide of British Columbia teen Amanda Todd after facing extortion and cyberbullying, the Ontario Lottery and Gaming Corporation, airport security, Scouts Canada, former CBC Radio host Jian Ghomeshi, the Communications Security Establishment, the asbestos industry, and Rob Ford, among countless other stories, have altered the country's political landscape. Not surprisingly, the show has the largest Twitter and Facebook followings of any Canadian current events program. *Enquête*, a weekly public affairs program on Ici Radio-Canada Télé, which broke story after story on corruption in the construction industry in Quebec in the early 2010s, has had a similar influence in that province.

CBC also has a formidable online presence. While the English- and French-language news sites, cbc.ca and ici.radio-canada.ca, sometimes seem like neon billboards that do little more than advertise network shows, they are among the country's top news hubs, featuring national and regional news and continual news updates. Together, they attracted some 18 million unique visitors a month by October 2017[34] and enjoyed a sizable lead in numbers of users over their Canadian private-sector news rivals. Unfortunately, as Barry Kiefl points out, the average Canadian spends less than three hours per year visiting cbc.ca out of the close to 400 hours that they spend online every year.[35] According to Wade Rowland's calculations, cbc.ca accounts for less than one-500th of the time that Canadians spend online.[36]

The CBC has pinned great hopes on its digital signature and sees its online presence as a gateway to the future. Yet in a battle against mega-sites such as Facebook, Google, WhatsApp, *BuzzFeed*, YouTube, and Instagram, among a host of others, it barely registers. Significantly, the CBC's digital ad revenue is little more than one-tenth of what it receives from television advertising, and the TV numbers are relatively small. The brutal reality is that when it comes to online advertising, two mega-companies, Google and Facebook, have formed what is, in effect, a duopoly, accounting for well over 70 per cent of the online revenue and leaving few leftovers on the table for others.

A Last Chance to Save the CBC

One important caveat in analyzing the CBC's position is that public broadcasting is in retreat almost everywhere. Reaching audiences has been as challenging and confounding to public service broadcasters in Europe and Asia as it has been for the CBC. The reasons for decreased audiences seem to be the same almost everywhere: less revenue due to falling support from governments, increased competition due to the explosion of channels on satellite and cable, and most critically, the digital onslaught that has scattered audiences and made platforms such as YouTube, Facebook, and Netflix the new gravitational fields of media power. Most critically, younger audiences are disappearing, leaving public broadcasters top heavy with older viewers and listeners.

There is a sharp difference, however, between public broadcasters in most other countries and the CBC. Despite a relative drop in audience numbers, public broadcasters such as the BBC in the United Kingdom, SVT in Sweden, ARD and ZDF in Germany, NRK in Norway, and RAI in Italy are still dominant players within their media spheres. The BBC, for instance, still has a 30 per cent audience share on TV and well over 50 per cent of the radio audience. It is the largest producer of dramatic programs in the world, with quite a number of its shows setting a global standard for dramatic excellence. It is also one of the world's great news organizations, with correspondents in dozens of countries, and is the leading source of online news in the United Kingdom. Not surprisingly, roughly a quarter of its revenues come from the global marketplace.

In Denmark, the two public broadcasting channels attract more than 60 per cent of the audience. In the Netherlands, the main public broadcaster has a market share of around 30 per cent, while in Italy, RAI captures more than a third of nightly viewers. The main public broadcasters in Germany together attract roughly 25 per cent of the audience. In Ireland, RTÉ One and RTÉ2, despite the fact that they face fierce competition from British broadcasters such as the BBC, ITV, and Sky, still command well over a quarter of the viewing audience, broadcast 15 of the 20 most watched programs in Ireland, and command 50 per cent of TV ad revenue.[37] In short, these are audience numbers that CBC executives can only dream of, numbers from a past that can never return.

Moreover, far from becoming what Trine Syvertsen has described as "analog museums," public broadcasters in quite a number of countries have been given the mandate and the resources needed to be "digital locomotives … spearheading the transition to information societies" and have thus led, rather than followed, the digital expansion.[38] In other words, these public broadcasters have become the fulcrum, the launching pad, for coordinated national efforts to steer digital change to the advantage of the state.

In later chapters, we chronicle the failure of the federal government's policies to come to terms with the challenges presented by media shock and the new attention economy and their many attempts to pretend the problems away. While the country's policy-makers have only recently woken up from a long sleep, and even this is questionable, there has been nothing less than a massive reordering and reinvention of the media world. The CBC is hardly alone in its fight for survival. Media consultant Ken Goldstein believes that almost all big-city newspapers and local TV stations will be gone by 2025.[39] While Goldstein's dire predictions may seem overdrawn, alarm bells have been ringing throughout the Canadian media system for quite some time. With Facebook becoming Canada's largest news distributor; with online revenue dominated by just two companies, Google and Facebook; with the growth of databases and the ability to target consumers; and with the threats posed to conventional broadcasters by streaming giants such as Amazon Prime, Netflix, and Disney; traditional media organizations in Canada are treading water. Over the last decade, their audiences and readerships have hemorrhaged, and they have had to cut back to the point where they are no longer recognizable. While French-language media organizations are not in as precarious a position as their English-language counterparts, the same forces of change endanger their existence as well—albeit, as we have mentioned above, at a slower rate.

In 2017, the Public Policy Forum, with the support of the government of Canada and a host of foundations and corporate sponsors, published a report entitled *The Shattered Mirror: News, Democracy and Trust in the Digital Age*. Focusing primarily on newspapers, the report found that the news media in Canada had hollowed out to the point where its very existence was in jeopardy. It quoted John Cruickshank, a former publisher of the *Toronto Star*, as observing that as investigative journalism and feature writing were becoming rarer and articles thinner and less complete, the news media was losing its ability to communicate with and reflect the country. To make matters worse, only half of Canadians were aware that the newspaper industry was crumbling, and fewer people knew that TV news was in sharp decline.

Not surprisingly perhaps, in 2017, Canadian broadcasters spent substantially less on news than they did on sports, a trend that is likely to accelerate.[40]

The facts on the ground are terrifying. According to the Canadian Media Guild, some 12,000 media jobs have vanished in the last few decades.[41] An American study found that nearly a quarter of newsroom jobs were lost between 2008 and 2017; this included almost half of all newspaper jobs.[42] In addition, local TV stations lost 20 per cent of their revenue between 2011 and 2015, with losses before interest and taxes hovering at one point at close to 10 per cent a year. The rate of loss slowed to 6.3 per cent in 2016–17.[43] This is a rate of attrition that simply can't be sustained.[44]

As private-sector news organizations move closer to the edge, they see themselves increasingly in conflict with the CBC.[45] As their employees face empty desks, voluntary buyouts, embarrassingly low salaries, and brutal workloads, they envy the subsidies that keep the CBC afloat and wonder why there is so little left on the plate for them. The public broadcaster is now under concerted attack from private media organizations, with the sense, perhaps on both sides, that there are only so many seats on the remaining lifeboats.

Complaints against the CBC have come from newspapers that have erected paywalls and charge readers for access to their sites. The problem is that their business strategy collapses if readers can still get news and opinion for free on cbc.ca or ici.radio-canada.ca. Similarly, critics complain that in opening its websites to exchanges of opinion, the CBC is encroaching on territory that once belonged almost solely to newspapers. A German law actually prohibits public broadcasters from providing the "press-like" online services offered by private media companies.[46]

While there are many factors that have conspired to weaken and humble the CBC, we believe that among the most critical factors is that it, and most other Canadian broadcasters, are unlikely to survive the endless kaleidoscope of the attention economy. While the CBC's many supporters argue that more money, more political support, doing away with advertising, a different governing structure, more local broadcasting, better management, or a winning streak that will come with hit shows might save the CBC, our argument is that the old playing field is gone forever. While all these changes might make a considerable difference in the short term, they are unlikely to be the keys to survival in the long run. What is needed is emergency-room surgery and not a change of prescriptions.

The End of the CBC? argues that the CBC is unlikely to survive for much longer against the assault of the new attention economy unless it takes on a different form. However essential the CBC's role was in the past, we cannot hold on to the fading world that once was. What we believe is needed is a basic rethinking of the role that public broadcasting can play in Canadian democracy and, indeed, in the country's continued survival. What we propose is a new ledger of responsibilities for the CBC. This will involve discarding much of what it does today in favour of playing a more concerted and focused role in only certain areas. Instead of the CBC being all things to all people, the "everything" broadcaster, we propose that it stick to doing a limited number of things very well. Our proposal is that the CBC needs to shed much of its old skin and become solely a news and current affairs organization dedicated to producing high-quality, dependable, and fair news and analysis in areas such as the economy, Canada and the world, health care, Canadian culture, and most important, the political life of the country. It needs to do investigative journalism, produce the accountability news that holds people and institutions responsible, and do the essential work of democracy. It needs to be the place where people turn to get news and perspectives about the country, its future, and the world that shapes it.

The crisis of news that we will describe in a later chapter and the crisis that is enveloping the CBC intersect and can be solved, or at least alleviated to some degree, by a single bold policy move.

This means that its relationship with the private broadcasters will have to be the reverse of what it is today. Whereas the CBC is now the broadcaster of first resort, expected under the Broadcasting Act to be all things to all people but deprived of the resources needed to achieve its mandate, private broadcasters need to shoulder some of the responsibilities that the CBC now carries.

The next chapter will describe the origins of public broadcasting in Canada and the forces that helped create it. As discussed earlier, the CBC was one of the great nation-building enterprises in Canadian history. The chapter will examine the factors that led to the CBC's current predicament, including the climactic changes in political ideology that have weighed against public service broadcasting, the broadcaster's hasty and ultimately devastating retreat from local broadcasting, and the CRTC's decisions to deny the CBC valuable footholds in the cable universe. The chapter will also chronicle the epidemic of political interference that made the CBC into a political football. In other words, we will discuss the factors that have led the CBC to the edge of a cliff.

The third chapter will take a close look at the slow erosion over time of the CBC's budgets and the effects that limited resources have had on CBC decision-making. The chapter will also explain how governments have largely ignored or pretended away the vast changes that have been transforming the media landscape. Looking back over time, government reports on the CBC's future have all seemed to look the same. They have promised change, often had dramatic titles, but done little to alter the landscape in any meaningful way. Whether because of the basic inertia that clogs the arteries of so much of what takes place in Ottawa or a real failure to imagine what seemed unimaginable as late as 2010, when streaming was technologically primitive, the government's record has been dismal.

The fourth chapter will describe the challenges faced by the CBC in the new attention economy. The main argument is that, with so many choices in drama, music, sports, and news, it is increasingly difficult for the CBC to reach its audience. As Virginia Heffernan has put it simply and poignantly, "Human discourse is now adapting to its new home."[47] That new home is Facebook, Google, WhatsApp, content farms, Apple devices, malware, mashups, memes, apps, hashtags, the hundreds of hours of videos uploaded to YouTube each minute, Instagram, Amazon Prime, facial recognition, the surveillance culture, and ESPN+, among a myriad of other devices and platforms. The chapter will explore three aspects of media shock, in particular: the power of new, multi-platform conglomerates such as Facebook, Amazon, and Bell Canada; the impact that Facebook and other social media are having on the future of news; and last, how Netflix and other streaming services have changed the culture of TV watching.

Chapter 5 will describe the effects of media shock and the new attention economy on news and sports programming, in particular. Wherever it turns, the CBC faces immense competition, and it has lost much of the audience that it once had. The argument is that if the CBC can no longer compete in these essential areas, then it is unlikely to survive.

In chapter 6, we explore the impact of the CBC's digital transformation, which is fracturing audiences, destroying the traditional methods of quantifying the breadth and impact of CBC programming, and forcing the broadcaster to chase every new innovation in search of listeners, viewers, and readers. Some of its new approaches, such as podcasting, may be working, but others clearly aren't as both audiences and the CBC's influence as a broadcaster quickly erode. We contend that while there is still time to make the tough choices that we believe will secure the CBC's future, the doors are closing.

A seventh chapter will examine the most recent attempts by the federal government to catch up to the digital universe that is unfolding before it. Recent policy documents seem to largely ignore or downplay any role that the CBC might play in spearheading change or any real plan to deal with the collapse of the Canadian media generally. Strangely, the government seems to have a Netflix policy rather than a media policy.

In the concluding chapter, we will return to the question of whether there is a place for public broadcasting in the new media age. We think there is, and we will propose a new CBC, with a new mandate and different priorities than the one that exists today. We will propose that the CBC jettison programming in areas where it can no longer compete or where, as in the case of drama, for instance, the National Film Board and the private sector can do the heavy lifting. By withdrawing to news and current affairs programming, the CBC will become larger by becoming smaller.

We also argue that, without these bold steps, public broadcasting in Canada is unlikely to survive.

Chapter 2

Lost Horizons

This chapter examines the factors that led to the creation of the CBC as well as the vision that guided it during, roughly, the first 50 years of its history. The chapter chronicles the period of broadcast scarcity, when the CBC was largely a singular and powerful voice able to shape Canadian life, to the cable era, when audiences were fractioning and the public broadcaster became just one voice amid a multiplicity of broadcast services. As Amanda Lotz has argued, cable changed the tide of broadcasting not only because it used a different broadcasting model, one based on appealing to narrow rather than broad interests, but also because it ushered in a new era of experimentation and excellence.[1] The chapter highlights a number of key decisions that led to the abandonment of the CBC's original vision: its forced retreat from local TV news, the decision by the CRTC to deprive the CBC of a significant place in the cable universe, and the complex and tangled relationships with governments that distorted its priorities and limited its budgets at the very time when it should have been expanding its services and ambitions. In short, this chapter is about the CBC's lost horizons and the role that successive federal governments have played in helping to undermine that original vision.

The State or the United States?

British scholar David Hendy argues that public service broadcasting was one of the most important political and social developments of the twentieth century.[2] This was certainly the case in Canada. Prime Minister Mackenzie King, who established the CBC in 1936 (after the failure of the first experiment, the Canadian Radio Broadcasting Commission, or CRBC, created in 1932 by Conservative Prime Minister R.B. Bennett), believed that through radio, "All of Canada ... became a single assemblage swayed by a common emotion…."[3] Politicians immediately understood the value of having their own voices heard from coast to coast, and governments intrinsically understood the power that came from being able to penetrate Canada's vast distances with the flip of a switch.

Part of the motivation for launching the CBC was that, by the end of the 1920s, the government was losing control over radio. Commercial broadcasting had become a free-for-all. Beginning with the issuing of the first radio licence in 1919 to the Marconi Wireless Telegraph Company of America for a station in Montreal, by 1927 the number of radio licences had mushroomed to close to 75. Most crucially, major American broadcasters such as RCA, Westinghouse, and CBS were making inroads into Canada's largest cities. But it was also that a hodgepodge of different models and interests had emerged. The Manitoba government owned two radio stations; the University of Alberta had initiated the first educational station in 1927; the Canadian National Railway had begun broadcasting on trains in 1923, eventually creating the first national radio network; and religious broadcasters were quick off the mark with their own stations. In addition, there was a growing wireless subculture of amateur operators and radio clubs. The federal government faced the need to bring order out of this chaos but also ensure that Canada's national interests were being served.

What worried the federal government the most was that radio reception was being compromised by powerful signals from south of the border and that Canadian commercial operators would serve only larger cities such as Montreal and Toronto, where they could make money, leaving vast parts of the country in silence, without radio. The main concern, however, was the fear that the Canadian radio system would become an adjunct of American broadcasting. Vancouver alone was bombarded by six high-powered American stations that dominated the airwaves from early morning to late at night. By 1930, an estimated 80 per cent of Canadian listeners preferred US programs.[4] To some Canadian nationalists, it seemed as if the game was lost almost as soon as it had begun.

What's interesting is that while so much has changed in the world of Canadian broadcasting, so much has remained the same—namely, the problem of how to enhance Canada's cultural sovereignty while living beside a financial, entertainment, and cultural behemoth. This primordial question in relation to Canadian broadcasting was famously posed by Graham Spry of the Canadian Radio League, a lobby group formed by prominent Canadians in 1930. The choice, he argued, was either "the state or the United States."[5] In other words, broadcasting could be

Canadian only if it were a public service. Canada could either choose a national public broadcaster that would be a "single glowing spirit of nationality" or allow Canadian broadcasting to become entwined with the American commercial system.

A.W. Johnson, the president of the CBC from 1975 to 1982 and one of the principal architects behind the creation of Canada's social programs, believed that countering the influence of the United States was at the core of the decision to found the CBC. As he put it, Canadians exposed to a relentless flood of American programming were "absorbing American interpretations of events ... soaking up the value system of the United States ... [they are] coming to expect Canadian traditions and institutions to look and behave as if they were American traditions and institutions."[6] In other words, the CBC's supporters saw public broadcasting as almost a life-or-death decision for the country.

The federal government faced another difficult dilemma, one that is largely forgotten today. Some religious broadcasters had begun using the airwaves to spread poisonous views about other religions and attacked governments as "the work of Satan."[7] Preventing radio from becoming a launching pad for religious warfare was a real, if a largely unstated, goal of public broadcasting.

The decision in favour of public broadcasting was made more likely by the fact that there was a successful example that could be emulated: the BBC, which had been created in 1924. It has to be remembered that, at the time, Canada was very much a British country. Most of its peoples had originally come from the British Isles; its institutions, including the monarchy and Parliament, were inherited from Britain; the Judicial Committee of the Privy Council (JCPC) in Britain would remain Canada's highest court until 1949; and Canada had yet to achieve foreign policy independence from Britain. The BBC was not just another broadcaster; it was for Canadians the premier example of what broadcasting should be.

The BBC was the product of forces that seemed to be coming together at the same time in British society but were also emerging in Canada. The BBC was, in part, a response to the fact that the old world was crumbling. There was the feeling not only that a better world had to be created following the carnage and suffering of World War I but also that new social forces had to be accommodated, including women's suffrage, the expansion of the vote to men who had previously been excluded because of a lack of economic means, and the rise of socialism. There were also the lingering vestiges of an age-old Victorian paternalism, which believed in helping the less fortunate. Consequently, underpinning the BBC's mission was a kind of social gospel. The BBC was all about improvement. It was all about educating the public and cultivating an appreciation among citizens for high culture, proper language, and more refined tastes.[8]

The first rough draft of what would become Canadian broadcasting policy emerged from the work of the Royal Commission on Radio Broadcasting, chaired by the then president of the Canadian Bank of Commerce, John Aird. In its 1929 report, the Aird Commission argued that broadcasting needed to be seen primarily as a public service. The report was strongly and resolutely in favour "of some form of public ownership, operation and control behind which is the national power and prestige of the whole public of the Dominion of Canada."[9]

Even with the Aird Commission's report in hand, creating the CBC was no easy matter considering the many challenges that the federal government faced. First, it wasn't clear that broadcasting was under federal jurisdiction. Both Quebec and New Brunswick were convinced that broadcasting was a provincial matter, and as mentioned previously, the government of Manitoba already owned radio stations. In June 1931, the Supreme Court of Canada ruled in a 3–2 decision that broadcasting was a federal responsibility. Although Quebec appealed the decision to the JCPC, the British court ruled in favour of the federal government, recognizing Ottawa's right to legislate under the "peace, order, and good government" clause of the British North America Act. Had the courts ruled in favour of provincial control, arguably, much of Canadian history would have been different.

Deciding how to finance the public broadcaster would take time to sort out. Until 1953, the CBC was funded by licence fees paid by Canadians when they bought radios and by a limited amount of advertising. Licence fees ended when the government calculated that the costs of TV broadcasting, which had just begun, would be far greater than they had been for radio. The licence fee was to be replaced by an annual appropriation from Parliament. One can argue that having backed away from the licence fee, Canadian governments would never again have a chance to fund broadcasting from what would be, in effect, an annual value-added tax on the purchases of communications and electronic equipment without incurring the public's wrath. The political price would be too great for any government. Some might argue that financing the CBC through an annual appropriation from Parliament also allowed governments to exercise greater political control over the public broadcaster.

In retrospect, the choice of an annual parliamentary appropriation was a poor one. The most successful public broadcasters, such as the BBC, SVT (Sweden), NRK (Norway), ARD and ZDF (Germany), NHK

(Japan), and RTÉ and TG4 (Ireland), among others, are funded by annual licence fees paid by TV owners. The fees are generally set by governments, with the size of the fees fixed for a certain length of time. In the United Kingdom, the fee is collected by the BBC itself, while in other countries, the fee is collected by the post office or comes as part of the electricity bill. The licence fee in the United Kingdom is approximately Can$260 a year, a figure that is close to five times higher than the amount that, according to a Nordicity study, at least, the Canadian federal government spends every year on public broadcasting in per capita terms.[10] Taking into account that the British population is almost twice as large as Canada's, it's no wonder that the BBC remains the force that it is. The fee is even higher in Germany, Denmark, Sweden, and Norway.

The licence fee means that public broadcasters are not dependent for their funding on the whims, vendettas, and sudden mood swings of politicians. Since they know years in advance, in approximate terms, what their funding will be, they have a longer planning window—a considerable advantage given that programs can take years to develop. Moreover, licence fees are set high enough so that, in most cases, public broadcasters do not have to carry advertising. This gives broadcasters much greater latitude to challenge governments and powerful corporations, take unpopular positions, and break controversial stories without fearing a backlash from politicians and advertisers. By contrast, the CBC has often found itself riding a financial roller coaster, not knowing what its budget will be from year to year. As will be discussed later, politicians on all sides have had few qualms about using financial purse strings to warn or punish the CBC. Moreover, since the CBC's funding is always measured against other priorities—health care, infrastructure, transportation, social services, equalization, housing, defence, university research, etc.—the CBC has often been an easy and sometimes popular place to cut. When it abandoned the licence fee, the federal government sowed the seeds of future troubles and constrained the future development and capabilities of public broadcasting.

After an awkward period in which the CBC alternated between French- and English-language programming during its broadcast day, prompting a growing outcry from listeners in Ontario, in particular, separate French- and English-language services were established in 1941. Despite having a common mandate, the CBC and SRC have, as discussed earlier, developed different programming content and styles, reflected different cultural and political values, and drifted into different orbits. Almost from the beginning, there was little contact, few crosswalks, and only small dollops of sharing. Patrick Watson, an on-air personality and producer who also served as chair of the CBC's board of directors, has described the frustration of those that tried to bring the two networks closer together: "The Corporation had tried a few times to get Francophone and Anglophone production teams to collaborate on a project, but it had never worked out, and our lunch group had concluded that it was because the orders had come down from on high rather than originating where program ideas best originate."[11]

It would not be long, however, before Radio-Canada began to play a major role in Quebec society. Indeed, one can argue that it became a major catalyst for the political change that would transform Quebec during the Quiet Revolution. During the 1950s, the broadcaster helped give life to Quebec's modern identity and publicized issues and criticisms of the roles played by the Catholic Church and the conservative Union Nationale governments, which were attempting to resist change. As historian Paul Rutherford has described the SRC's impact:

> Radio-Canada offered to the Québecois a concrete, visible expression of their own unique places, past and present, and ways. "Television in Québec," Susan Mann Trofimenkoff has observed, "magnified the tiny world of the Laurentian village, a lower town Québec, or a local hockey arena into a provincial possession." Its newscasts and its public affairs shows plus the many, many features and documentaries swiftly created a novel means of focusing attention on the activities and concerns of the province.... This drama didn't so much create as perpetuate and update a cluster of symbols that gave definition and meaning to the community.[12]

More controversial is the fact that Radio-Canada undoubtedly gave voice to and, to some degree, sparked the ardent nationalism that would dominate Quebec politics through the FLQ crisis (which involved the Montreal kidnapping of British trade commissioner James Cross and the kidnapping and murder of Quebec Labour Minister Pierre Laporte by members of the separatist Front de libération du Québec), the constitutional battles of the late 1980s and 1990s over the Meech Lake and Charlottetown Accords, and the referendums on Quebec sovereignty, held in 1980 and in 1995. While some critics believe that Radio-Canada actively promoted Quebec sovereignty, it is more accurate to say that Radio-Canada would have lost all credibility if it had not reported on the great debates about Quebec's sovereignty that were swirling at the time and had not discussed the sovereigntist option that many Quebecers were actively considering.[13] Nonetheless, the resentments and turmoil that characterized Quebec politics for much of the late 20th century, and the painful struggle for national unity that divided Canadians during the 1980s and 1990s, raise critical questions about the role that a public broadcaster devoted to promoting Canadian identity should play when the very existence of the country is at stake.

In the early years, CBC programming was mostly musical recordings and concerts. It wasn't long, however, before hockey broadcasts regularly attracted audiences of a million listeners or more, a staggering number for a country that had a little more than 11 million people in 1935. Another popular program was *The National Farm Radio Forum*, which began broadcasting in 1941. It reached hundreds of thousands of listeners, spawned some 1,600 discussion groups, had its own weekly newspaper, and produced mountains of study materials.[14] News broadcasts of the results of the 1935 federal election, the Moose River mine disaster in April 1936, and the Royal Tour in June 1939 (the first two events broadcast by the CBC's predecessor, the CRBC) brought the country together in ways that had never occurred before. Perhaps the main reason for this transformation was the unique nature of radio listening. Unlike TV, which provides viewers with visual images, radio forces people to visualize the scenes that are being described, to fill in the empty spaces with their own imaginations. Moreover, radio is a profoundly intimate medium because listeners feel that they are being spoken to directly. It wasn't long before Canadian political leaders such as Alberta premiers "Bible" Bill Aberhart and Ernest Manning, Maurice Duplessis in Quebec, Mitch Hepburn in Ontario, and Tommy Douglas in Saskatchewan were using the airwaves to build popular followings.

Radio's great power in those early days came from the fact that people rarely listened to it alone. Listening to the radio was an experience that brought not only families but also, often, friends and neighbours together. TV, on the other hand, is a medium that is occasionally communal within families, but much less frequently than in radio's golden era.

The most galvanizing changes to the CBC occurred during World War II. While roughly a third of Canadian families had radio-receiving sets in 1931, three-quarters had them by 1940, and historian Mary Vipond estimates that "almost all" had them by 1950.[15] Before the war, the CBC's news bulletins consisted largely of reports culled from the Canadian Press wire service, the British United Press, and newspapers. The public's insatiable demand for war news forced the CBC to establish its own news service in 1941. It wasn't long before the daily *News Bulletin* was supplemented by a *News Roundup* entirely devoted to war news. CBC journalists such as Matthew Halton, Peter Stursberg, and Marcel Ouimet were with Canadian troops on the front lines in Europe and reported with remarkable eloquence on the lives and struggles of the soldiers that they covered. The CBC was part of the national war machine, with reporters and, indeed, senior managers integrated into Ottawa's larger propaganda strategy. Journalists wore army uniforms, referred to the Canadian Army as "we," and festooned their battlefield reports with tributes to the bravery of the Canadian soldiers.[16] Its repertoire of patriotic shows included *The Army Sings, They Fly for Freedom, Men at War,* and *The Life Line Holds.*

Censorship was a grim reality. First, because of the sensibilities of the times, news bulletins, even those that did not involve the war, would tone down or omit descriptions of murders and violent crimes as well as divorces and suicides so as not to shock or offend listeners. Most crucially, bitter defeats such as the August 1942 raid on Dieppe were masked to appear as victories, and the CBC avoided giving any publicity to those who opposed conscription for overseas service, even as the country held a divisive referendum on the issue in 1942.

TV broadcasting began in Canada in September 1952, when CBC stations were launched in Montreal and Toronto. By 1956, there were 26 TV stations in the country. As had been the case during the radio era, Canadian TV had to compete with the popularity of US border stations and American shows and stars. Those programs, such as *I Love Lucy, The Honeymooners, The Ed Sullivan Show, Jack Benny, Gunsmoke, The Lone Ranger, Lassie,* and *The Twilight Zone,* among others, became the mainstays of the Canadian viewing experience and, indeed, set the standard against which Canadian TV would be judged. The pull of American stations was magnified by the fact that American local stations aired National Football League (NFL), National Basketball Association (NBA), and Major League Baseball games. Some Canadians formed attachments to sports franchises such as the New York Giants, Cleveland Browns, or Boston Celtics that would last a lifetime.

Despite living under Hollywood's long shadow, Canadian TV was able to take some important first steps. Not surprisingly, *Hockey Night in Canada* (HNIC) immediately became CBC TV's most popular program. With the help of talented announcers such as Foster Hewitt and Danny Gallivan in English and René Lécavalier on Radio-Canada, audiences soared to 3.5 million in English and 2 million in French by the early 1960s. *HNIC* was the only program that routinely attracted larger audiences than American shows. As a side note, Danny Gallivan became famous for reinventing the English language with his patented descriptions of "cannonading" blasts, "larcenous" puck-stealing, and "spinarama" moves.

The CBC took advantage of hockey's popularity by scheduling its main variety show, *Juliette,* immediately after the hockey games on

Saturday night. The show became the main stage for emerging Canadian talent for almost a generation. Comedians Johnny Wayne and Frank Shuster, who had established themselves as radio stars during the war, provided off-the-wall and zany comedy skits on Sunday nights. CBC's entry into the world of quiz shows, a genre that dominated television in its early years, was *Front Page Challenge*, on which celebrities were questioned about events in the news or from history by a panel that included journalists Pierre Berton and Betty Kennedy and the cantankerous Gordon Sinclair. In addition to airing these popular shows, the CBC attempted to bring high culture to Canadians in the form of operas, plays, and ballet. Although they fit the old BBC model of what a public broadcaster should be doing, these shows had limited audience appeal.

On the French-language side, the largest audiences were for hockey, musical variety programs such as *Café des Artistes* and *Music Hall*, and *La famille Plouffe*, a show that achieved almost legendary status in Quebec popular culture. The most revolutionary and controversial program was undoubtedly the current affairs show *Point de mire*, hosted by René Lévesque, who would later become the face of the Quebec sovereignty movement as leader of the Parti Québécois and premier of Quebec. With great panache, Lévesque would provide viewers with a weekly tour of the world's trouble spots using photos, maps, and an old-fashioned blackboard. The show's popularity proved that public broadcasting could be an extraordinary vehicle for public education and a catalyst for social change.

Aside from the major American networks and several independent Canadian stations, CBC TV faced little domestic competition until CTV arrived on the scene in 1961. CTV was originally set up as a loose consortium of independent broadcasters that teamed up to buy American shows and broadcast sports events such as the Grey Cup. TVA, the largest French-language private network, founded in 1971, was, like CTV, originally designed as a cooperative owned and operated by its affiliate stations. CTV and TVA would later be taken over by media giants Bell Media and Québecor, respectively.

With the growth of private broadcasting, the federal government created the Board of Broadcast Governors in 1958 to take over the regulation of the broadcast system from the CBC. Until then, the public broadcaster had been in the uncomfortable position of being both broadcaster and regulator. A successor agency, the Canadian Radio and Television Commission, was established in 1968, with its jurisdiction extended to include telecommunications in 1976. The CRTC now adjudicates the entire communications highway, from local TV stations to

Internet and wireless service providers, from the amount of Canadian content that must be aired on radio and TV stations each day to the rates charged to customers by telecoms and mobile phone companies. As we will argue in the next section, the CRTC has rarely been a friend to the CBC. In fact, the regulator was instrumental in tilting the playing field in favour of private broadcasters.

Shifting Sands

The Broadcasting Act of 1991, which still remains in place as the basic constitution of the Canadian media system, states explicitly that "where any conflict arises between the objectives of the Corporation [referring to the CBC] ... and the interests of any other broadcasting undertaking ... it shall be resolved in the public interest...." Arguably, the CRTC has turned a blind eye to enforcing this part of the Broadcasting Act. In a relatively short period of time, beginning in the 1960s and 1970s, the broadcasting system went from one dominated by the CBC, with private broadcasters serving as an adjunct to the public broadcaster, to one dominated by private broadcasters, with the CBC treated as an afterthought. The CBC's fall from grace can be attributed to a number of factors, some of which lie at the CRTC's doorstep.

Critics argue that the CRTC has adhered to an unwritten "national champions" policy that has protected the four main media giants—now Bell Canada, Corus, Rogers, and Québecor—so that they can be profitable enough to withstand the onslaught of Hollywood competition and produce top-quality Canadian programming. In other words, the hope is that they will be too large and too prosperous to fail. For many years, until the recent digital explosion and the advent of the new attention economy, they were also "too protected" to fail. The protective cover given to the privates has taken a number of forms. When TV and radio licences come up for renewal, approval appears to be automatic despite the fact that broadcasters have sometimes slipped wide of the mark in living up to their previous licence-renewal promises.

As mentioned in the previous chapter, private broadcasters have also benefited from simultaneous substitution—the blocking of American advertising on shows that are being aired on US and Canadian stations at the same time—which gives Canadian advertisers a "right of way" in reaching Canadian viewers when they are watching popular American shows. Most critically, private companies have been allowed to expand into the cable universe, first as owners of specialty channels and then later as providers of cable services, while the CBC was largely shut out of what would be broadcasting's most lucrative treasure trove. In addition, while the CBC benefited from the subsidy cushion for Canadian content that the federal government gave to independent producers, the ultimate effect of the program was to redirect to private broadcasters money that might otherwise have gone to the CBC.

Another unstated policy was "Canadianization through Americanization." The idea was to allow private broadcasters, with fewer Canadian-content obligations than the CBC, to load up on Hollywood shows in prime time. This gave the private networks larger audiences, and because they benefited from well-known American stars and the much larger publicity campaigns of US TV networks, they could attract larger audiences and advertising and generate greater profits than would be the case if they were making a Canadian show from scratch. In fact, the basic math was that the cost of buying hit American shows off the shelf from Hollywood (at prices far below the original costs of production, to begin with) vastly reduced risk and almost always guaranteed a profit. The logic was that Americanization would make private broadcasters more profitable and, in doing so, give them the resources needed to spend more money on Canadian content. To put it mildly, the spending splurge never came.

There are several reasons why this reversal of fortune in favour of private broadcasters took place. Perhaps most critical was that the nature of the debate about the role of government changed dramatically in the 1980s and 1990s. During the 1960s and 1970s, when Keynesian economics was in full bloom, there was a supposed government solution for every societal problem, and government expenditures ballooned accordingly in Ottawa, in Quebec City, at Queen's Park in Ontario, and even in Alberta. The ideological temperature changed considerably in the 1980s and 1990s. An anti-state, conservative tide led by Ronald Reagan in the United States, Margaret Thatcher in the United Kingdom, and Mike Harris in Ontario and Ralph Klein in Alberta would lead to a reduction in the size of governments and to the wholesale privatization of Crown corporations. Conservative critics saw governments as bloated, wasteful, and tangled in red tape, while the private sector was trumpeted as being more efficient and grounded in reality. That led to the belief that governments should be run as if they were a business. Added to this were concerns during this time that the federal and other governments were hitting a debt wall from which they might never recover.

Florian Sauvageau, who is a former Radio-Canada on-air personality and is regarded as among Canada's leading media experts, has pointed out that the new wave of ideological politics wasn't only about downsizing the CBC. It was also about transforming it from the inside. He recalls the changes that occurred within the CBC citadel:

> The chorus of praise for the private sector was accepted. Subcontracting and independent production prevailed. Artisans were evicted from the institution. They certainly worked less quickly than those in the private sector, but they had developed a "culture" of quality and meticulous work which reflected the values and mission of public service. For example, in the case of television dramas …, authors—great writers whose talent and depth had produced original television dramas—were replaced by the factory of Hollywood-style made-for TV movies, where flashy technique often conceals the superficiality of production-line scripts and pathetically poor writing. Concern for quality is time consuming but imperative in creation if we are to surpass the McWorld (fast food, fast drama) of mere consumption.[17]

According to Sauvageau, the end-result of this "privatization from within" was that the CBC became less distinctive, less original, and ultimately, less worth watching.

In recounting this history, the authors don't necessarily accept the moral dichotomy that Sauvageau draws between the public and private sectors. Much of what appears on private networks is compelling and popular with audiences, and much of what appears on the CBC has missed the mark. In fact, one of the key points that we will make later in this volume is that cable channels such as HBO, Showtime, and FX have led the way in terms of experimentation and innovation and that the threat to the CBC comes from the excellence of their programming rather than its poor quality. Canadian private broadcasters were decidedly not part of this broadcasting revolution.

The question is whether this tilt in the direction of private broadcasters was so great that it jeopardized the very existence of the CBC. Why did federal politicians see such little value in public broadcasting? The answer is not a pretty one. We will argue later in this volume that benign neglect and an inability to see the future that was unfolding before them played a key role in sidelining the CBC and much of the Canadian media until it was too late. Cabinet ministers, senior bureaucrats, the CRTC, and the broadcasters themselves all played a part in missing the storm that was surrounding them everywhere. There is, however, another more disturbing part to the story—namely, that the petty grievances, suspicions, and animosity of political leaders of all stripes led them to see the CBC as a critic that needed to be silenced rather than an institution that was designed to be a watchdog on behalf of citizens and a necessary counterweight to their power.

The academic literature on the relationship between political leaders and the press suggests that it is rooted in both symbiosis and conflict.[18] Governments and political parties need journalists in order to reach the public with their messages. In fact, political success rests on the ability of political leaders to set the media agenda, to have journalists accept their issues and proposed solutions as being the most critical to the country rather than those of their opponents. At the same time, politicians can never fully control the agenda. In fact, almost all political campaigns go through a "crisis," in which they lose control of the agenda to their opponents and the press.[19] The relationship between politicians and journalists is particularly acute and desperate when political leaders believe that the public broadcaster has a sacred duty to carry their messages and become frustrated that they are paying the bills for the very organization that is attacking them. Not surprisingly, virtually every prime minister since the 1950s has been at war with the CBC in some form or other.

Perhaps the most vitriolic relationship was between the CBC and Progressive Conservative John Diefenbaker (1957–62). At various points, Diefenbaker believed that the CBC had been infiltrated by communists, was subverting public morality with indecent programming, and was trying to sabotage his TV appearances. He talked openly about "cleaning up the CBC," froze the salary of CBC President Alphonse Ouimet, and attempted to cancel *Preview Commentary*, an editorial opinion segment that aired on CBC Radio immediately after the daily 8:00 a.m. news. Since *Preview Commentary*'s goal was to stir debate, it frequently featured opinion pieces that were critical of the government. Notoriously vain and thin-skinned, Diefenbaker thought that the show had become a political dagger at his throat. *Preview Commentary* was eventually saved from the scrap heap after Frank Peers, the head of CBC public affairs programming, and more than 30 other producers threatened to resign. Diefenbaker also helped (by some accounts) secure a TV licence for his friend and political ally John Bassett. Bassett's Toronto station, CFTO, came to dominate the Toronto television market and was one of the linchpins in the creation of CTV.

Liberal Lester Pearson (1963–68), often seen as the very embodiment of cautious diplomacy, seemed to lose all caution when it came to the CBC.[20] When members of the government were mercilessly criticized, cross-examined, and lampooned on the CBC's landmark show *This Hour Has Seven Days*, Pearson declared war on it. After the Prime Minister's Office (PMO) launched an "inquiry" into the show, the network capitulated and took it off the air.[21] The very fact that a prime minister could orchestrate the cancelling of a highly popular TV show because it mocked his government is chilling. It is as if Donald Trump were able to order *Saturday Night Live* off the air because he didn't like the way he was portrayed in its comedy sketches.

Pierre Trudeau (1968–79, 1980–84), who often appeared as a commentator on Radio-Canada before he entered politics, was equally frenzied in his attacks on the public broadcaster. Believing that Radio-Canada had been taken over by separatists and had succumbed to "a sickness of spirit," he threatened to shut the network down. As he once suggested, "We will put a lid on the place … we will close up shop. Let them not think we won't do it. If need be, we can produce programs, and if not, we will show people Chinese or Japanese vases instead of the nonsense they dish out…."[22] While one might dismiss this as just wishful and wistful thinking, watching and listening to criticisms of the country from Quebec nationalists on Radio-Canada was difficult and almost unendurable for federal politicians.

Jean Chrétien (1993–2003) was, if anything, even more antagonistic than Pearson and Trudeau. Chrétien saw Radio-Canada as a *"boîte à séparatistes"* (nest of separatists), and he blamed the network for treating the sovereigntist side with "kid gloves" during the Quebec referendum of 1995 and for ignoring or downplaying his speeches.[23] Like Trudeau, he mused privately about closing the corporation, while engaging in a long war of budgetary attrition with the public broadcaster. In a crucial period when the CBC needed to expand into the specialty universe, it had few, if any, friends at the Cabinet table.

The lack of friends included Brian Mulroney (1984–93). Obsessively conscious of media coverage, Mulroney used all the tricks of the media trade to win favourable coverage. He would flatter, charm, leak to, confide in, befriend, and punish journalists to bring them onboard. This would mean phone calls to reporters, anchors, and editors, complaining about how he had been "short-changed" in a particular story, as well as phone calls for the same purpose to the president of the CBC.[24] On at least two occasions, the CBC found itself in his crosshairs for coverage that he had found particularly threatening or unnerving. He was deeply upset by the CBC's coverage of the Meech Lake constitutional accord and especially by criticisms of his actions made by Elly Alboim, the English-language network's Parliamentary Bureau chief, in a speech at the University of Calgary. Alboim had suggested that Mulroney's "sole motivation" in launching the Meech Lake initiative had been "to establish that he could do in Québec what Pierre Trudeau could not." When, according to rumours at the time, CBC President Gérard Veilleux tried to fire Alboim, he faced open rebellion from the corporation's leading journalists, including Peter Mansbridge.[25]

Then there was *The Valour and the Horror*, a three-part docudrama that aired in 1992 that was critical of Canada's military during World War II and especially the bombing of German civilian populations. The series inflamed passions among veterans and created a national dust storm that Mulroney couldn't ignore. He set a Senate subcommittee loose to hold hearings about the series' accuracy, a spectacle of public shaming rarely seen in Canadian media history.[26]

Mulroney also stacked the CBC's board of directors with staunch Conservatives, some of whom were bitter critics of the CBC. One appointee, economist John Crispo, believed that the CBC was anti-American and far too liberal in its values, and he admitted that cuts made to the CBC's parliamentary appropriation in 1990 were both "a down payment and a warning."[27]

Patrick Watson recounts how he felt when Mulroney was considering appointing him CBC president as part of a coup against then president Pierre Juneau: "The idea of a government firing a CBC president for political reasons was anathema, a crude and dangerous violation of the arm's-length relationship. Roy [Faibish] was doing his best to persuade me that it was in the interests of the survival of the Corporation, since as long as Juneau was there a Tory government would consider the CBC to be the enemy. But it was still a political move; it was against all the best traditions of public broadcasting in Canada, including the express intent of the Broadcasting Act, which protects the CBC president from government interference."[28] Mulroney appointed Watson chair of the CBC's board of directors instead.

In the next chapter, we will discuss Stephen Harper's efforts to downsize and arguably slowly strangle the CBC. His contempt was more than palpable. He would avoid interviews with the CBC, and virtually all his appointments to the CBC's board of directors were people who had donated to Conservative campaigns.

There are two points to consider. First, criticism of politicians by journalists from private-sector organizations was often as biting and negative as that by journalists from the CBC. Political leaders, however, seemed particularly aggrieved when criticism came from news or entertainment shows that they saw as being funded by their own governments. Second, public broadcasters in most countries have, from time to time, found themselves in the crosshairs of politicians. It is worth noting, however, that in countries where public broadcasters are funded by licence fees, they don't seem to be as nakedly exposed to the whims and vindictiveness of politicians as the CBC has been. Although this culture of public hostility at the very top of the Canadian political pyramid may have diminished with Justin Trudeau, for much of the corporation's history Canadian political leaders have been unable to resist trying to turn the public broadcaster into a partisan political weapon. When criticized, their reaction has been to try to control it using threats and intimidation. In the heat of partisan battle, none thought to redefine its role by giving it the re-envisioned mandate and financial tools that it needed to survive the approaching tsunami of digital change.

Three Points of No Return

While we pointed out in the introduction that the CBC has either retreated from or been displaced from a whole series of media worlds, we believe that three turning points were particularly significant. The culprit in each case was a budgetary squeeze that forced the CBC to retreat from major parts of broadcasting. These were its decision to sacrifice much of local broadcasting, its almost wholesale exclusion from cable, and its inability to hold on to *Hockey Night in Canada*. While we argue in a later chapter that the loss of hockey may have saved the CBC in some ways, taken together these three dramatic moves on the chessboard altered the nature and future of public broadcasting.

Perhaps the most decisive move occurred in 1990, when the CBC made a strategic decision to save the "centre" at the expense of local news. It did so by closing down supper-hour news shows in "secondary" cities and merging them into regional broadcasts. In Calgary and Edmonton, for instance, the two supper-hour news shows merged into what became a bewildering and hare-brained compilation of Calgary and Edmonton news stories. Audiences in both cities fled the regional-news-hour show as if it were a burning building and migrated in ever-increasing numbers to private stations. In fact, despite several attempts over the years by the CBC to re-establish its local news shows in Calgary and Edmonton as well as in other markets, audiences never returned in appreciable numbers. With few resources and little more than skeleton crews, the shows in all too many key centres regularly run last in the ratings.

While saving the centre at the expense of the regions may have made sense to desperate CBC managers dealing with budget cuts, with the end of local programming and cuts to local news a large number of people in communities across Canada thought that the CBC had become little more than a mirror of distant Toronto. The ripple effects were staggering. In one fell swoop, the CBC not only cut itself off from audiences in burgeoning cities such as Calgary and Edmonton, but gave CTV and then Global the vital footholds that they needed to build large local followings. Moreover, as supper-hour news shows helped divert audiences into the private broadcasters' prime time schedules, the impact was reflected in smaller CBC audience numbers throughout the schedule for years to come.

We noted in the introduction that the amount of money that the corporation spends in Toronto and, of course, in Quebec far exceeds what it spends in the so-called regions. The CBC is in this sense asymmetrical. It has a different size and shape in different parts of the country: looming large in some areas and barely noticeable in others.

Another crucial decision, this time made by the CRTC, was to largely exclude the CBC from the specialty channel universe. While the CBC did receive licences for CBC Newsworld (later renamed CBC News Network) and subsequently Radio-Canada's Ici RDI as well as a handful of niche channels, the CRTC turned down applications for eight other specialty channels.[29] The problem was twofold. In the late 1980s and 1990s, the CBC was too financially threadbare to be able to bid for new specialty channels in sports, children's programming, music, or history, and the CRTC had signalled that it wasn't going to award these licences to the CBC, in any case. Specialty channels were especially lucrative because, unlike conventional channels, which were solely dependent on advertising, cable channels had two financial pillars: advertising and fees paid to broadcasters from cable subscribers to each of the specialty channels. In some cases, these fees amounted to 90 per cent of a channel's revenue, depending on how channels were bundled in the packages that cable companies sold to consumers, thereby almost freeing the companies from the need for advertising. Had the CBC been awarded key licences in sports or music, for instance, it would have enjoyed additional revenue streams, greater economies of scale by being able to air shows multiple times, and key gateways to new and younger audiences.

Most important, as the University of Michigan's Amanda Lotz has pointed out, cable became the fulcrum for media change because broadcasters had to carve out clear brands and offer distinctive programming in order to survive.[30] While much of early cable was a wasteland of reruns, old movies, and oddball programming, in later years cable channels such as HBO, Showtime, and A&E would launch original shows that broke the mould, such as *Sex in the City, The Wire, Breaking Bad, Mad Men*, and *Game of Thrones*. Forced to stand on the sidelines, the CBC watched the revolution pass it by.

The critical point to remember is that, in other countries, the cable explosion took place on the public as well as the private side.

Governments elsewhere seemed to understand that public broadcasting could not be successful without the additional oxygen created by expansion into the cable frontier. Simply put, cable brought a new attention economy, which necessitated an increased presence in order to be seen and heard. There seemed to be no such realization in Canada, at least, for securing a place for public broadcasting.

Another point to consider is that being a cable provider (broadcasting distribution undertaking, in CRTC parlance), as separate from owning specialty channels, would become enormously lucrative. Private players would become both. Companies such as Rogers, Bell Canada Enterprises (BCE), Shaw, and Québecor would eventually become cable and satellite providers as well as owners of specialty channels. Most critically, as the media giants invested heavily in digital infrastructure beginning in the late 1990s, cable companies would become major links to and providers of Internet services as well as online broadcasting.

University of Calgary professor Gregory Taylor has argued that CBC News Network is "hitched to a broken system" of cable subscription fees and advertising that can't be sustained; after all, the cost of subscribing to it online is $7 a month, while a subscription to Netflix was $9 in 2017. (That was modified in late 2018 with the launch of Gem, the CBC's own streaming service. While the basic service is free, it includes advertising; Gem Premium is available for $4.99 a month, which includes News Network and advertising-free content for all other programming.) There are innovative ways to strengthen the news service, including making it freely accessible online and using BBC iPlayer to reach audiences in the same way that the BBC does.[31] While Taylor is no doubt right, there is no cure for the fact that the CRTC, by awarding cable licences almost exclusively to the private sector, severely curtailed the CBC's growth and future possibilities.

A third turning point, a third point of no return, was the loss of *Hockey Night in Canada*. While we will discuss sports broadcasting at much greater length in a later chapter, the handover of *HNIC* to Rogers, while beneficial to the CBC in some ways, was a stark and unforgiving defeat in other ways. Without the finances needed to compete for the rights to broadcast NHL games, the CBC not only fell into the degrading position of broadcasting hockey games that were owned, operated, and staffed by Rogers, with Rogers getting all the advertising revenue, but also lost its place at the centre of hockey culture, a place that it had occupied for the previous 70 years.

This pattern of loss now seems to be repeating itself. The same lack of resources that prevented the CBC from competing effectively in local

TV news, taking advantage of the cable revolution, and winning the rights to broadcast NHL games also prevented it from acting decisively in the face of media shock. The key point is that the CBC had begun to lose its place in the broadcasting solar system long before the digital revolution hit with full force. We now believe that, with the advent of the new attention economy, the CBC is again at a crossroads.

Chapter 3

The Politics of Resentment and Neglect

The titles of the CBC's annual reports dating back to 2004–05 say a lot about how the corporation would like Canadians to view public broadcasting.

What Is a Public Space?
Striking the Right Balance
How Do You Cope with Constant Change—Stay in Front of It
Challenging, Informative, Entertaining, Canadian
Great Successes, Greater Challenges
The Old Rules No Longer Apply: Reshaping Canadian Public Broadcasting
Yours in Every Way
More Canada, More Regional, More Digital
Challenging the Status Quo
Going the Distance—Sochi 2014
Canadian Content, Regional Advances and Digital Enhancements
Canada's Public Space
Celebrating Canadian Culture
Canada 150!

It is the image of a corporation responding positively at the forefront of change. But that's not the CBC reality that most Canadians have lived for the past quarter-century. Controversy, cutbacks, upheaval, and attempts at consolidation are the constants, not conquering change.

During that time, neither Liberal nor Conservative governments ever stated clearly what they wanted the public broadcaster to be or do. Whichever party was in power viewed the CBC through the lens of what it could do for them or, more frequently, how it could hurt them. As Wade Rowland has noted, public broadcasting "may make for a culturally literate, well-informed public but it does nothing to make managing politics easier for those in power."[1] We argue that vanity and petty grudges seemed to determine the fate of Canadian broadcasting—despite warnings that a new, tumultuous era was dawning.

More critically, governments seemed unwilling and unable to respond to the massive changes brought about by the digital revolution. While reports and policy documents paid lip service to change, the Canadian government chose to do almost nothing, to sit on its hands, as change ripped through and overturned the media system. In a relatively brief span of 20 years, mega-platforms such as Google, Amazon, and Facebook came to dominate almost all aspects of media, audiences splintered to the point where every user had the capacity to create their own individualized media ecosystem, and cellphones became the main screen and, indeed, the constant companion for most Canadians. Canadian governments, however, seemed to be stuck in another age. Instead of using the public broadcaster to be the agent of change, as other countries have attempted to do, they relegated the CBC to the sidelines, where it increasingly became a spectator rather than a participant in the new world that was unfolding.

This chapter will first describe the budget cuts that diminished and hamstrung the CBC during a period when new opportunities were emerging. We will then discuss the attitudes toward public broadcasting exhibited by the Chrétien and then Harper governments. In the case of both governments, resentment and suspicion, combined with a lack of imagination, seemed to be the main policy instincts.

Last, we will look at the stream of reports by government, committees, and the CRTC that while pointing to problems, led to very little in terms of policy changes by government. Report after report would pile up—leading nowhere.

Sidelining the CBC

Since 1993, the CBC has paid for the suspicion and ambivalence of governments and political leaders many times over. That year, the Liberals under Jean Chrétien returned to office after nine years of Progressive Conservative government under Brian Mulroney and, briefly, Kim Campbell. As discussed in chapter 2, these PC governments were hostile to almost everything associated with public broadcasting. They regularly appointed CBC board members with no media background or track record as supporters of public broadcasting and, in some cases, seemed to relish appointing people who were ardent and even bitter critics. The Mulroney government cut the CBC's parliamentary appropriation to $857 million in 1986 from $904 million the year before. But funding did grow in the following years, as Canadians debated free trade with the United States and the Meech Lake and Charlottetown constitutional packages. As the country found itself at a crossroads, the public's attention to politics and feelings of nationalism were at a high point. By 1993, the CBC's parliamentary appropriation was back up to $1.097 billion.

Once in office late that year, the Liberals initially continued with modest increases until financial support peaked at $1.17 billion in 1995–96. In 1996–97, major spending cuts announced in the Liberals' February 1995 budget began to bite. The CBC's appropriation fell a bit, to $997.1 million in 1997, before plunging to $806.4 in 1998, then climbing slightly to $879.2 million by 2000.

With respect to budget cuts, the math is beyond dispute. Simply put, the size of the annual appropriation from Parliament is barely larger, in current dollars, today than it was in 1991. The grant fell slightly, from $1.078 billion in 1991 to $1.036 billion in 2014.[2] In 2015, the newly elected Justin Trudeau government agreed to inject an additional $75 million

Figure 1. CBC Parliamentary Appropriation, 1980–2018 (Thousands of dollars)

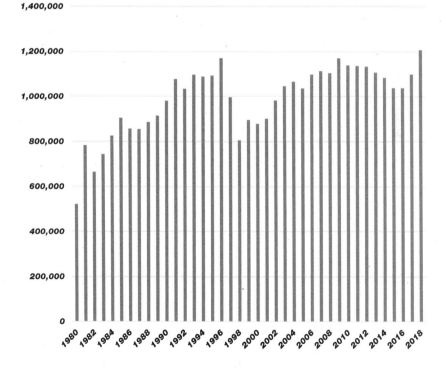

into the CBC's budget in 2016 and $150 million per year for the four years after that, allowing it to hit $1.208 billion in 2017–18. This was a significant move because it lessened the year-by-year announcement of the annual appropriations that had constrained long-term planning at the CBC for more than 20 years because the corporation never knew what its budget would be more than a year in advance. Given that the time between the original planning and the first airing of a new TV show could be two to three years or longer, the lack of more secure funding into the future meant that the CBC was always scrambling to address short-term needs rather than taking bold steps such as launching new cable channels or commissioning more expensive, serialized TV programs such as AMC and HBO and then Netflix had begun to produce. It is hard to escape the conclusion that political leaders of all stripes chose to leave the CBC dangling because it gave them greater control.

Even with this boost in funding under Trudeau, if one takes the ravages of inflation into account, the parliamentary appropriation has dropped by close to one-half since 1991. Cumulative inflation during that period, as calculated by the Bank of Canada, has been 64.5 per cent, so that $100 in 1991 would be worth $164.5 in 2019.[3] Had the appropriation increased only by inflation every year, it would be $2 billion in 2019.

But the cost of programming has also increased during that time for, as the saying goes, a billion dollars isn't what it used to be. In 2018, a single *Spiderman*, *Star Wars*, or *Avengers* movie cost well over Can$200 million to produce. Netflix's mega-drama *The Crown* cost close to Can$150 million to make. A single season of *Orange Is the New Black* cost more than Can$60 million. This means that the CBC's entire current parliamentary allocation could pay for just three or four blockbuster movies and a handful of Netflix shows. The CBC's parliamentary allocation can't come close to matching the more than US$8 billion that Netflix alone spent on original dramas in 2018. Amazon Studios spent close to US$6 billion on original productions in 2019—some five times the CBC's entire budget. Time Warner did the same.

A 2016 study conducted by the Nordicity Group, which compared per capita expenditures on public broadcasting in 18 Western countries, found that Canada ranked a dismal third from the bottom, ahead of only New Zealand and the United States.[4] Topping the list were the Scandinavian countries, Switzerland, Germany, and the United Kingdom, whose expenditures on public broadcasting were three or four times greater than Canada's. While the Nordicity study is controversial because critics argue that the figures don't include the subsidies given to independent producers or the consequences of tax policies that direct Canadian companies to advertise in Canada rather than the United States, both of which benefit the CBC, Canada's low ranking is unmistakable.

The reasons for the budgetary savaging are multi-faceted. First, as mentioned in the previous chapter, the nature of the debate around the role of government changed dramatically in the 1980s and 1990s. Keynesian economics had been in full bloom in the immediate post-war era, and governments were expected to intervene and invest in the economy and provide solutions to societal problems. The CBC was part of this shift to big government. The ideological climate shifted considerably during these decades, however, with a conservative, anti-government tide led by Ronald Reagan in the United States, Margaret Thatcher in the United Kingdom, and provincially by Mike Harris and Ralph Klein in Canada. Governments were now seen "as part of the problem"—to use Ronald Reagan's famous description. They had become too bloated and self-serving to be effective, and they interfered with the marketplace, which was expected to provide new ideas and products as well as jobs.

In this environment, government-supported broadcasting seemed to be a ghost from the past. In addition, as mentioned in the previous chapter, the CBC had to compete for budgetary dollars against other federal government priorities, which the public often viewed as being more immediate and essential: health care, infrastructure, and social programs such as pensions, post-secondary education, etc. And most critically perhaps in the often bitter and hotly contested world of federal-provincial relations, transfers to the provinces—with few strings attached—meant that the provinces rather than the federal government had most of the money.

Added to this were concerns in the late 1980s and 1990s about the federal and other governments hitting a debt ceiling from which they might never recover. In this frenzied environment, the CBC became an easy target for both the Mulroney Conservatives and the Chrétien Liberals. But it needs to be pointed out that cuts to other government programs, including transfers to the provinces, were equally draconian.

It's also the case that many Canadians have an ambivalent attitude toward the CBC. While support for public broadcasting is generally high, a sizable minority of Canadians doesn't support public broadcasting, either because it sees the corporation as not reflecting its values or because it just doesn't watch or listen to the CBC and thinks that it shouldn't have to pay for other people's media. John Meisel, a former head of the CRTC and one of Canada's early political scientists, has identified what may be a more acute and sensitive problem. According to him, Canadian culture, including the CBC, is mostly supported by what he described as an "elite" audience: older, better educated, and more likely to be women. At the same time, the majority of Canadians are awash in American mass entertainment and culture. His devastating claim is that Canadian culture is the minority culture in Canada. The majority culture is American. Meisel believes that this painful reality has produced a self-fulfilling prophecy. Since most Canadians spend almost all their time watching American TV and films, listening to American music, and visiting American online platforms, they see little value in Canadian cultural institutions, including the CBC, and are reluctant to support those institutions with their tax dollars.[5] As a consequence of this chronic underfunding, Canadian cultural institutions are weaker and less attractive and, thus, less able to complete. The cycle is rarely broken.

Between 1994 and 2000, the Liberals cut the CBC's budget by close to $400 million. Perhaps not coincidentally, the public broadcaster was in almost non-stop battles with a prime minister whom *Toronto Star* television columnist Antonia Zerbisias described as having had a "hate-hate relationship with CBC for years."[6] That came largely from Jean Chrétien's history with Radio-Canada and his ambivalence, which we discussed in the previous chapter, about the role that it had played during the 1995 Quebec referendum on sovereignty. Chrétien thought that while his pronouncements were often ignored, Lucien Bouchard, the leader of the pro-sovereignty Oui side, received uncritical, even adoring, coverage. Zerbisias noted other examples of coverage that drove Chrétien's animosity—some going back to the beginning of his pursuit of the Liberal leadership. In 1984, Chrétien arrived for a panel discussion on *The Journal*, the CBC's then flagship, prime-time current affairs program, the night Pierre Trudeau stepped down as prime minister, "only to discover that he would follow a lengthy documentary on John Turner's likely success as leader."[7] Chrétien did lose the leadership to Turner later that year and, as is the case with so many politicians, maintained a long memory, always believing that the public broadcaster had it in for him.

Chrétien's Liberals initially appeared supportive of the public broadcaster, and it seemed that the end of the Conservatives would mean better days for the CBC. Chrétien agreed to appear at annual town hall meetings broadcast by the CBC instead of the more traditional end-of-the year interview between the prime minister and journalists. With the new government enjoying the traditional post-election honeymoon, the initial town hall at the end of 1993 was a success for the

Liberals. The following year went well too, but the atmosphere got a little tense in December 1995, weeks after the Quebec referendum, at the third nationally televised, end-of-year session between Chrétien and selected members of the public. As *Toronto Star* columnist Rosemary Speirs noted, "Because of the Québec crisis the show that aired last night was a much harsher experience for the Prime Minister with questioners suggesting he ought to resign if he can do no better for Canada than that narrow federalist win on Oct. 30."

What happened the following year led Chrétien to refuse to participate in any future CBC town hall meetings. Audience member Joanne Savoie from Montreal asked Chrétien why he had broken a major 1993 campaign commitment to eliminate the federal 7 per cent goods and services tax (GST). Chrétien responded, "Did you read the Red Book [Liberal campaign platform document] on that?" he responded. "That's not what we said on that. You should have read it." He added, "We never said in the Red Book or directly that it was to be scrapped," suggesting that the Liberals had only proposed simplifying the tax. Ms. Savoie stood her ground. "I didn't hear simpler. I heard scrapped," she replied. "From whom?" a visibly upset Chrétien responded. "From you on television," she shot back.

She was right. On several occasions between 1992 and 1994, Chrétien had promised that a Liberal government would abolish the tax, as the media and opposition parties in Parliament were happy to point out immediately following the town hall. It was a disastrous performance for Chrétien, as Speirs described:

> Viewers saw a Prime Minister on the defensive, glossing over Québec's discontents and the broken GST promise and replying to human problems of joblessness and uncertainty with statistics. His questioners, picked by the CBC to speak on the job worries at the top of the public agenda, showed no mercy. One even asked the PM, "how could you sleep at night?"[8]

The same day, the normally Liberal-friendly *Toronto Star* headlined an editorial "The Prime Minister Is Lying," stating, "In front of a room full of ordinary Canadians and thousands more watching on television, the Prime Minister of Canada told a lie. Not a fib, not a prevarication, not a disingenuous remark—a brazen outright lie."[9]

It got worse for Chrétien in the following days as he tried to defend his comments against attacks on his credibility. He attempted to turn the tables on the CBC in his next interview with the broadcaster. On 21 December 1996, on CBC Radio's weekly political program, *The House*, with CBC journalist Jason Moscovitz, Chrétien noted that there were

200 people in the audience but only 13 questioners, suggesting that he was set up. He also accused Moscovitz on several occasions of feeling guilty and defensive about the event, which Moscovitz forcefully denied.

As the *Globe and Mail*'s Edward Greenspon reported, "Asked directly about reports that aides in his office believe the event was rigged, Mr. Chrétien replied, 'You ask CBC, not me.' The prime minister was also critical of how newscasts used the exchange with Ms. Savoie, stating 'there's always cut and editing and so on to suit the purpose, not necessarily to have the whole truth,' adding to Moscovitz, 'You're a professional. Do you think you acted professionally? Only you can answer that.'"[10]

It was not the end-of-year send-off that the Liberals had hoped would propel them into the election year of 1997. But it was also not Chrétien's last clash with the CBC over how it reported on his government. Safely returned to office with a second majority, Chrétien met with leaders of the Asia-Pacific nations at the annual Asia-Pacific Economic Cooperation (APEC) summit, hosted in November that year by Canada in Vancouver. During that summit, the Royal Canadian Mounted Police (RCMP) pepper-sprayed demonstrators rallying against Indonesian President Suharto over human rights violations in his country. "The problem is that they tried to jump over the fence and I'm telling you that was not acceptable," Chrétien told a news conference, responding to questions about aggressive police actions after first noting, in vintage Chrétien style, that "for me, pepper is something I put on my plate."[11] He indicated that the demonstrators were given a location where they could peacefully hold their protest. When a few tried to jump a fence, Chrétien said, the RCMP pepper-sprayed them in the face to preserve order.

It sounded good, but it wasn't true. Television newscasts showed the RCMP walking up to and pepper-spraying protesters who were calmly sitting on the road. There was no fence-jumping by demonstrators or peaceful, initial attempt by the police to move the protestors. That confrontation, and Chrétien's flippant dismissal of the RCMP's actions, began a year-long investigation of the police force's role in security at the APEC conference by the CBC's Vancouver reporter, Terry Milewski. Much of his efforts concentrated on trying to determine whether the PMO had directed the RCMP's actions on that day, including the pepper-spraying. Correspondence between Milewski and an activist involved with the demonstrators, when later made public, led to an extraordinary complaint by Peter Donolo, Chrétien's director of communications, to the CBC's ombudsman, claiming anti-Chrétien bias on

Milewski's part. Donolo stated that, in that email exchange with protesters, Milewski "referred to the federal government as 'the Forces of Darkness,' [that Milewski] waged 'a concerted campaign' to 'milk' the APEC issue and 'promote[d] a one-sided account while working secretly with an interested party on the matter.'"[12]

In response, Milewski removed himself from the story while CBC conducted an internal investigation. Before the ombudsman ruled, CBC news management responded to the complaint in a letter to Donolo, stating, "We strongly reject that Mr. Milewski was blind to all the facts emerging and argue his on-air reportage speaks to issues of fairness and balance." The letter added, "Mr. Milewski is an aggressive reporter. He also has a sense of humour which comes across not only in conversations but also in the tone of his personal correspondence."[13]

The PMO wasn't laughing, and neither was Milewski. He refused to be interviewed by the CBC's ombudsman, former journalist David Bazay, based on a previous ruling by Bazay in an unrelated matter that the reporter considered unjustified. So the issue was referred to Radio-Canada's ombudsman, Marcel Pépin, who ultimately found Milewski innocent of all the PMO's allegations. While Pépin was considering the issue, the CBC suspended Milewski for 15 days after he wrote about the controversy in the *Globe and Mail*, accusing the PMO of trying to intimidate journalists. It wasn't until March 2000 that the CBC and Milewski settled, with the broadcaster reinstating his pay for his suspensions during the previous two years.

Perhaps it's too much to infer that vanity and revenge played such a large part in deciding the CBC's budget and, indeed, in limiting and shrinking the broadcaster—at the very time that public broadcasters in other countries were being given the means to expand. But there can be little doubt that during the years that the PMO and CBC were battling, the public broadcaster was shrinking. In fact, it faced a perfect storm. At the same time that its parliamentary appropriation was being cut and its employee numbers were being chopped, its audience was fracturing in different directions as a result of the explosion of cable services. While funding recovered slightly after 2000 and during Paul Martin's two years as prime minister, the CBC was bludgeoned again during Stephen Harper's years in office. At the risk of repeating ourselves, we need to point out that the budget in current dollars in 2019 is barely what it was in 1996 without calculating inflation.

To help the CBC cope with the cuts, shortly after 2000 the Liberals began increasing the appropriation by $60 million a year, describing it as additional, non-recurring funding for programming initiatives. That ended under the Conservatives in 2012, when the $60 million was rolled into a reduced parliamentary appropriation. As mentioned above, as long as that money lasted, CBC management was in the dark every year about whether it had that extra $60 million—about 6 per cent of its appropriation—until the federal government released its budget. That additional annual amount was an effective tool to restrain the public broadcaster, keeping it beholden to the government of the day and presumably receptive to any hints about how the CBC's programming could support the government's political agenda. The unstated message was blunt: step out of line, and that money could disappear. If it were suddenly removed, there would need to be immediate cuts to programming, operations, and staffing.

While the parliamentary appropriation grew slowly after the cuts of the late 1990s, what didn't grow after the mid-1990s was employee numbers. In 1987, the CBC had 10,778 employees, about two-thirds of them male. Just more than a decade later, in March 1998, as the Chrétien cuts bit, employment was down to 7,316 full-time and contract employees—a decline of one-third in the decade. The cutting occurred despite the first hint, in the corporation's 1996–97 annual report, of the explosion that would come from the newly emerging digital economy—the disruptive force that, over the next 20 years, would undermine all assumptions about the role, future, and even existence of public broadcasting. A paragraph in that CBC annual report simply noted:

> In response to Canadians' growing desire to inform and entertain themselves through new media, the Corporation has developed a strategy and dedicated modest but recurring funds to develop a mandate for the new media era. With 32 per cent of Canadian households equipped with computers and half of them with modems (1996), there is definitely market potential for new media, the World Wide Web in particular, in Canada. The CBC's mandate is to provide Canadians with "a wide range of programming that informs, enlightens and entertains." New media and the digitization of production open up a multitude of new opportunities for the Corporation.[14]

This turned out to be truer than the authors could imagine at the time, but the CBC would have to adjust to the upheaval with fewer employees on the front lines. Over the following decade and a half, there was virtually no growth in employee numbers in the face of what seemed to be almost annual rounds of layoffs, so that by 2016–17, total employment was 7,555. A year later, on 31 March 2018, CBC had 7,444 employees.

Table 1. CBC Employees through the Decades, from Selected Annual Reports

	1987*	1998*	2011*	2016–17	2017–18
Total employment	10,778	7,316	8,729	7,555	7,444
Full-time	10,211	6,728	7,285	6,626	6,377
Temporary	539	588	465	313	432
Contract	28		979	616	635
Among full-time employees					
Female	3,287			3,240	3,640
Indigenous				132	151
People with disabilities				159	181
Visible minorities				735	908

*Complete data not broken out in annual reports before 2016–17.

While employment was virtually static, far more devastating was the fact that the CBC was losing its audience. Audience shock did not come all at once, but in slow, agonizing twists in the wind. These losses were, in part, a by-product of the rise of specialty channels. As discussed in chapter 2, through the 1990s and into the 2000s there was a dramatic growth in the number of channels that Canadians could purchase on cable and satellite systems, but the CBC's repeated efforts to become a major player in the cable universe made little impression on the federal government or the CRTC. In 1980, the CBC announced plans to create CBC-2 and Télé-2 as second television services, featuring regional programs for a national audience, experimental projects, and foreign programming, all originating in CBC studios for distribution by cable companies. CBC proposed that all cable systems would be required to carry the channels, and the public broadcaster would receive the monthly subscriber fees. But the cable companies objected, and the CRTC responded by rejecting the idea. Lobbying from the private sector played a role as the privates didn't want the CBC cutting into the specialty-channel cash cow that they wanted for themselves.[15]

As mentioned in chapter 2, that denial of CBC specialty-channel applications by the CRTC occurred eight times during the subsequent two decades. In many ways, those decisions crippled the CBC because its private-sector, over-the air network competitors were the same companies that were scooping up specialty-channel licences. As Wade Rowland commented,

While television audiences were badly fragmented by the many new viewing options, owners of both conventional television stations and specialty cable channels were able to reassemble those audiences over multiple platforms ... not only was the public broadcaster denied access to the fountains of revenue represented by cable pass-through fees, but it had to cope with exponentially increasing competition for audiences in virtually all genres of programming in this new highly fragmented universe.[16]

Stated differently, one can argue that the CBC's mandate to cover programming in all areas—children's programming, current affairs, comedy, sports, music—was being dismantled and handed over almost lock, stock, and barrel to the private sector.

In the cable universe, the CBC's all-news channel, CBC Newsworld, was up against not just CTV News Channel but also the venerated BBC, CNN, CNBC, Fox News, Al Jazeera English, and a host of foreign-language broadcasters. Where CBC Sports once had the sports lane almost all to itself, it now faced TSN and Rogers Sportsnet, both of which would mushroom to include multiple channels. In the 2000s, new genres of channels popped up dealing with subjects such as home renovations, food, outdoor living, real estate, and personal makeovers. They were just some of the new and popular arrivals, many financed by CRTC-mandated funds to support Canadian program production. All helped further fracture audiences to the detriment of the CBC. In what would eventually become a 400-channel universe if you were fully loaded, the public broadcaster had little to entice its audience to stay.

As described in the next chapter, the biggest blow has come from big-budget, serialized, dramatic productions produced by innovative broadcasters such as HBO, Showtime, AMC, FX, and A&E. Where the CBC once had to compete against only Global, CTV, and the main US networks, its relatively small stable of dramatic shows now have to compete for attention against an estimated 500 scripted, original series that air every year. Many, if not most, of these shows have higher budgets and noticeably greater production values than shows aired on the CBC. In the age of *peak TV*, the CBC would quickly become a bit player—as would Canadian private broadcasters.

The audience backslide has had major consequences. Smaller audiences mean less advertising revenue to offset stagnant or declining annual parliamentary appropriations. To add to the misery, because the CBC has only a small toehold in the specialty channel universe, it receives little in subscriber fees. Specialty channels, mostly owned by the parent companies of CBC's over-the-air competitors, Global and CTV, receive up to 90 per cent of their revenue from subscribers. The CBC has had no such buffer against declining revenues.

Interestingly, Canadian consumers pay more in cable or satellite fees than the amounts that consumers pay in licence fees to support public broadcasting in other countries. Instead of this money going to support public broadcasting, as it would in most other democratic countries, the "fees" that Canadian consumers pay go to Canadian specialty channels. Canadian shows that are aired on specialty channels also benefit from a host of government subsidies. The question is whether the trade-off between a thriving CBC and a kaleidoscope of specialty channels has worked for Canadians. Critics argue that most of what appears on cable and satellite are cheaply made, bargain-basement shows that are mostly knock-offs of foreign TV programs. Programs such as *Top Chef Canada, Canada's Top Model, Storage Wars Canada, Four Weddings Canada, Big Brother Canada, First Dates Canada, Real Housewives of Vancouver*, etc. have contributed, they contend, little if anything to Canadian culture and have rarely if ever reached for excellence in any way. While this criticism ignores the existence of broadcasters such as APTN, CBC News Network, and Ici RDI as well as the many sports, history, and music shows that have resonated with Canadians, it can be argued that the Canadian broadcasting system as a whole has lost in this reshuffling of the financial deck.

Cass Sunstein believes that the proliferation that came with cable was the first blow in the bludgeoning of "general interest intermediaries"—those TV networks, newspapers, and magazines that provided shared cultural experiences for society at large.[17] Once the floodgates opened to hyper-customized and personalized media experiences, to a universe in which people were able to construct their own media ecosystems, many of these general-interest intermediaries would fall by the wayside, although at least some of them with the potential benefit of a global audience—such as the *New York Times* and the BBC—would grow in influence. Sunstein argues that general-interest intermediaries are essential for a healthy, democratic society because "increased fragmentation prevents true (and valuable) information from spreading as much as it should" and because being exposed to a wide range of views and options makes freedom possible.[18]

What is critical to understand is that in a country where shared experiences are the lifeblood of Canadian identity and national unity, the survival of the CBC and other general intermediaries seemed to stir little interest. As media scholar Elihu Katz famously asked in 1996 at the time of the cable explosion:

Why are governments contributing to the erosion of nation-states and national cultures? Why don't they see that more leads to less to insignificance ... to endless distraction, to the atomization and evacuation of public space? Why don't they see that national identity and citizen participation are compromised? Why don't they realize that they're contributing directly to the erosion of the enormous potential which television has to enlighten and unite populations into the fold of national culture?[19]

Interestingly, countries like Germany, the United Kingdom, Japan, and Sweden were able to split the difference: maintaining strong public broadcasters at the same time that they were expanding viewing options for their publics. Eventually, those same public broadcasters would play leading roles in the digital revolutions in their societies. The solution in Canada seemed to be to make the CBC smaller.

Different Government, Same Story

The Liberals weren't the only ones using public broadcasting as a whipping boy. The Conservatives under Stephen Harper consistently attacked the CBC in their fundraising letters, arguing that the public broadcaster (joined by the rest of the media) was staffed by Liberal sympathizers who were determined to misrepresent and disparage everything the Conservatives said and did. Like Liberal complaints about Radio-Canada's sovereigntist leanings, Conservative paranoia about public broadcasting also had a long history, most recently in the party's antecedent, the western Canada–based Reform Party. It came to Ottawa in the 1990s as self-proclaimed outsiders and regularly disparaged the media as part of a Liberal-dominated Ottawa elite out of touch with the concerns of "hard-working" Canadians.

That theme became party doctrine when Harper, a former Reformer who had merged that party's successor, the Canadian Alliance, with the Progressive Conservative Party, formed a minority government under the Conservative Party banner in 2006. Over the following decade, the prime minister and the party consistently attacked the CBC. It was part of a broader communications strategy that portrayed the media in general and the CBC specifically as enemies of the government, following a pattern established by US President Richard Nixon's vice-president, Spiro Agnew, in the early 1970s and almost four decades before President Donald Trump adopted the same tactics.

A former head of news and current affairs for Radio-Canada, Alain Saulnier, in his book entitled *Losing Our Voice*, described how the Harper government mimicked the Chrétien Liberals in their attempt to manipulate, control, and weaken the CBC.[20] Arguably, the Conservatives went even further than the Liberals did in what amounted to an unrelenting war against public broadcasting. Conservatives, including the prime minister, would occasionally boycott interviews with the CBC, refuse to acknowledge its reporters at news conferences, and voice their displeasure over coverage with senior executives, including the corporation's president, Hubert Lacroix, in what can only be described as an attempt to intimidate them. At the same time, the Conservatives took every opportunity to express their disdain for the public broadcaster. That emerged most consistently in communications between the Conservative Party and its supporters and potential donors. A 2008 fundraising letter, for instance, from its campaign director, Doug Finlay, described the CBC as being "anti-Conservative,"[21] calling the public broadcaster "a vested interest" ally of the Liberal Party.[22] The letter was sent at the same time that the House of Commons Standing Committee on Canadian Heritage was in the midst of a year-long study looking at future directions for the corporation.

At the start of October 2008, two weeks before that year's general election, the lobbying group Friends of Canadian Broadcasting released a public opinion survey that it had commissioned, concluding that almost two-thirds of Canadians thought that money spent by the federal government on the CBC was a good use of taxpayers' dollars. The Friends' survey, conducted by Ottawa pollster Nik Nanos, came directly from a question posed in a previous Finlay Conservative fundraising letter: "The CBC cost taxpayers over $1.1 billion per year. Do you think this is a good use of taxpayers' dollars or a bad use of taxpayers' dollars?" Finlay's letter promised that he would forward the results to Prime Minister Harper. The Friends' poll results found little support for the Conservatives' war against the CBC. In fact, of those who identified themselves as Conservatives, 68 per cent said they favoured either the status quo or an increase in funding.[23] Nonetheless, CBC bashing became a regular theme in fundraising pitches throughout the Harper years. The Conservatives were aware, no doubt, that there was a difference between the anti-CBC views of many of their core supporters and those of ordinary Canadians and even those who were likely Conservative voters, but not party members, who were generally supportive of public broadcasting.

Re-elected with another minority in 2008, the Harper government continued to attack the CBC, and remarkably, it showed no interest in updating the Broadcasting Act or reimagining the Canadian media system even as Facebook, Twitter, and Google were beginning to change the face of media power and overturn the structures and assumptions of the entire media industry. It was as if the government's obsession with the CBC had become a cause in and of itself and a substitute for any serious thinking about policy issues.

It wasn't long before the government was again on the attack. The new heritage minister, James Moore, publicly chastised CBC executives about excessive expense account spending. Moore, a British Columbia MP, contrasted overspending by management with the CBC's decision to shut down the CBC symphony orchestra, based in Vancouver. The CBC responded by cutting executive travel and hospitality spending. Later that month, Finance Minister Jim Flaherty tossed the CBC's name into a discussion about the possible sale of selected federal assets to balance the budget.[24] Flaherty may have been floating the idea to see how it would fly.

In early 2009, Moore mused about the CBC becoming a public broadcaster like the Public Broadcasting System (PBS) in the United States, relying on viewer fundraising to replace advertising. (He didn't mention that PBS also receives significant revenue for some programming from corporate sponsors, whose support is noted only at the start and end of programs rather than in advertisements interspersed throughout a program.) Moore even suggested that his government might find some one-time money to help the CBC make the transformation.[25] The Conservatives seemed to be suggesting that an ever-shrinking CBC could be made to shrink even further.

The Conservatives also used complaints to the CBC's ombudsman to try to undermine the public broadcaster's credibility. One flurry of complaints in 2010 came in response to comments made by Frank Graves of Ekos Research, one of four polling firms under contract to the CBC at the time. A *Globe and Mail* column quoted Graves as advising the Liberals and their then leader, Michael Ignatieff, to open "a culture war" against the Conservatives. Conservative Party President John Walsh quickly labelled that a sign that the CBC used partisan pollsters and that the Ekos poll results lacked credibility. At the same time, Finlay cranked up the fundraising machine, again complaining in a new letter to donors about the hostile interests lined up against the Conservatives and asking recipients to complain to the CBC's ombudsman about the broadcaster's use of partisan pollsters such as Graves.[26]

After receiving 800 complaints, CBC Ombudsman Vince Carlin concluded that Graves wasn't a CBC journalist, hadn't violated the corporation's Journalistic Standards and Practices, and hadn't made partisan statements in any of the work he had done for the corporation. More generally, Carlin, a long-time journalist and journalism educator before becoming ombudsman, stated a truism: that "every government—Trudeau, Clark, Turner, Mulroney, Campbell, Chrétien, Martin and, now, Harper—has seen the press, and the CBC specifically, as 'hostile' to their intentions."[27]

In fact, a postscript to this attack emerged several years later in yet another Conservative fundraising letter. A letter to Harper written by then CBC chairman Timothy Casgrain in the midst of the Graves controversy became public in 2014 through an access-to-information request. As *National Post* columnist Andrew Coyne wrote, "Casgrain complained that the Conservatives' ongoing campaign against the CBC carried out through 'fund-raising letters' and 'in talking points distributed to government Conservative MPs' was 'unfounded in fact' and 'wilfully destructive of an asset of the Crown.' To the extent that these attacks might be used to influence the CBC's reporting, Casgrain accused the Conservatives of threatening the broadcaster's independence."[28]

The re-emergence of the Casgrain letter gave the Conservatives still further ammunition. They circulated yet another fundraising letter, claiming that the CBC "tried to prevent the Party from writing and emailing to Canadians about CBC bias." But as the party's director of political operations valiantly told potential donors in his letter, "We refused to be strong-armed.... The CBC cannot dictate what we Conservatives can say or do."[29] Of course, it was all complete baloney, but it may have been effective in the party's continuing campaign to portray the public broadcaster as the enemy.

The Conservatives went one step further in a fundraising letter in early 2014, which, in conspiratorial tones, informed supporters that "over 80 per cent of Canadian media is owned by a cartel of just five corporations." The newspaper industry, in particular, is "largely controlled by a small number of individual or corporate owners, which often own the television networks." This media convergence, noted Fred DeLorey, the Conservative Party's director of political operations, meant that the Conservatives couldn't get a fair shake from anyone in the media, adding that media oligopoly "has greatly complicated our Conservative Party efforts to present the unfiltered facts and foundations behind our policies."[30] For this one fundraising letter, in other words, it wasn't just the CBC; all media organizations were out to get the Conservatives. Only frequent and large financial contributions from party supporters could apparently overcome this conspiracy.

The Conservatives weren't the only ones that used the public broadcaster as a fundraising tool. The Liberals, trying to rebuild after their disastrous 2011 campaign, which gave Harper a majority government and reduced the Liberals to third place, circulated a "Hands off the CBC" letter, asking for donations to help fight for public broadcasting. This was a bit of a surprising turn given the Liberals' previous track record of vilifying the CBC and strangling its budget. Interim party leader Bob Rae claimed that the Conservatives were going to cut the federal deficit by further reducing the CBC's parliamentary appropriation (a move that did, in fact, occur). "The Liberal Party of Canada recognizes the profound importance of the CBC's role in our society," he wrote. "We will fight to ensure our national broadcaster receives the support and resources it needs to continue to do its vital job."[31] The Liberals' sudden about-face may have been the result of poll numbers that showed widespread support for public broadcasting as well as a grassroots campaign by the Friends of Canadian Broadcasting to mobilize support for the CBC in closely contested ridings.

One means of control exercised by the Harper Conservatives was to stack the CBC's board of directors with party loyalists. According to the Friends of Canadian Broadcasting, by 2013 it appeared that almost every single member of the board was a Conservative partisan, and quite a number had contributed money to the party.[32] According to Alain Saulnier, the ultimate lever of control, however, was selecting Hubert Lacroix as CBC president in 2008 and reappointing him in 2012. With the exception of appearing as a commentator on sports shows, Lacroix had no experience in the broadcast industry. He consulted frequently with the PMO and seemed to comply with the government in giving them detailed information about where and how he would make budget cuts. Worse still, according to at least one source, in talking to senior staff he would warn against anti-Conservative bias in stories and make it clear that there was a relationship between news coverage and future budget funding.[33] Instead of having a president who would represent the interests of the corporation to the government, it seemed to be the very opposite: Lacroix would represent the interests of the government to the corporation.

The government's ability to flood the zone with partisan appointments as well as choose a compliant board chair and president screamed for reform. Needless to say, the Harper government did nothing to reform the process, preferring instead to weaponize its power of appointment.

To tighten the vice even further, during this period Québecor, the dominant player in the Quebec media, used its many media outlets to launch a series of attacks against the CBC. At the time, in addition to owning TVA and Le Journal de Montréal and Le Journal de Québec, the media behemoth owned the Conservative-leaning Sun Media chain and a short-lived specialty news channel, Sun TV. Its reporters made extensive use of federal access-to-information legislation, filing hundreds of requests over several years for details of CBC spending, from which it hoped to generate stories it could inflate into scandals.

In late 2011, the CBC responded, alleging that Québecor had received more than $500 million of public money through television production grants and subsidy programs as well as gaining bandwidth during the previous three years for its Vidéotron wireless subsidiary without competitive (and usually expensive) bidding. While much of the CBC's response was based on questionable assumptions and math, it demonstrated that the corporation would react if sufficiently provoked—at least by corporate competitors.

The Conservatives' relentless campaign to reduce the size of the CBC continued in 2014–15 with a new round of government cutbacks. In that year, the CBC released a new, five-year plan entitled *A Space for Us All*, one of the centrepieces of which was a plan to eliminate an additional 2,500 jobs by 2020.[34] Saulnier described the plan as "a mess, a glib, lazy piece of work with senior management surfing on a few bits of modish jargon."[35] Moreover, "It was not a strategic plan at all, but a mass layoff plan disguised as a strategic plan."[36]

Surprisingly, in the midst of the 2015 federal election campaign, Prime Minister Harper asserted that the CBC had financial problems not because of government cuts but because fewer people watched or listened to it. "There aren't cuts," he told a Quebec City radio audience in an apparent rewriting of history. "The reason is the loss of [CBC's] audience. It's a problem for the CBC to fix."[37]

The Conservatives continued to flog the CBC during the 2015 election as Harper refused to appear in any televised leaders' debates organized or broadcast by the CBC. The party's antagonism toward public broadcasting continued after it moved to the opposition side of the House of Commons. Several candidates in the party's 2016–17 leadership race, including winner Andrew Scheer, suggested at various points dismantling all or part of the CBC, something the party never had the courage to attempt during its decade in power. The heat of a leadership race, where attacking the public broadcaster is the equivalent of throwing meat to pit bulls, is not the same as facing the larger Canadian public, where the CBC retains substantial support.[38]

Nonetheless, the damage inflicted by both the Chrétien and Harper governments remains. Subterfuge, brutal cuts, and inaction in the face of mammoth change has left the CBC deeply weakened. While an injection of new funding by Justin Trudeau's government has provided a life preserver of sorts, that life preserver might not be enough. The CBC can no longer compete in most genres, and the action has largely gone elsewhere. Its very weakness, its inability to do anything in a big way, has become a main argument for eliminating it. The weaker it becomes, the less Canadians would miss it or even notice that it was gone. As

Saulnier has argued, "The Conservatives behaved like owners who wilfully neglect their property and leave it to deteriorate until the only
option is to tear it down."[39]

Reports Collecting Dust

While both Liberal and Conservative governments share responsibility for diminishing and sidelining the CBC, neither party ever had
to pay a penalty at the ballot box for their actions. Although public
opinion polls repeatedly showed that a large majority of Canadians
believed that the CBC played an important role in strengthening Canadian culture and identity and supported either an increase in or the
maintaining of current levels of funding for public broadcasting, neither Chrétien nor Harper seemed to be impressed. Arguably, public
support for the CBC created a "red line" that no government wished
to cross—namely, eliminating the CBC. While Canadians may have
supported the CBC in the abstract, the budget cuts never kicked up a
dust storm of any kind from the public. The outcry never came. In fact,
according to one commentator who observed the last round of Harper
cuts, "The new cutbacks, far from raising public indignation, are taking place in quasi-general indifference. The public broadcaster is being
bled before our eyes and we are watching it in prolonged death throes
without making a fuss about it. As if maintaining high-quality television and radio programming and national and international news
and current affairs coverage worthy of the name had not the slightest
importance to us."[40]

One can also argue that no one ever won or lost their seats in the
House of Commons because of the positions they took on the CBC
because the role of the public broadcaster never emerged as a major
election issue—at least, compared to a host of other issues. There is one
major caveat, however, to the argument that politicians had a free pass
from the public. The lobby group Friends of Canadian Broadcasting,
run at the time by Ian Morrison, launched an information campaign in
key ridings across the country. Its members were encouraged to raise
CBC issues at all-candidates meetings and display Friends' campaign
lawn signs and buttons. At the very least, it became more difficult for
candidates to ignore what was happening to the CBC.

What is remarkable, however, is the degree to which governments
of all stripes remained oblivious to the vast changes brought about by
the digital revolution and the arrival of a new attention economy. In
specific terms, parliamentarians produced a series of reports and studies that made recommendations, starting in the mid-1990s, to reshape

public broadcasting for the digital age. While there was a great deal of brave talk, and ideas came and went, nothing ever changed. Some studies generated short-term debate between those who wanted more money and support for public broadcasting and those who wanted to kill it. In the end, these reports did little more than gather dust.

Soon after coming to power in 1993, the Chrétien government asked Pierre Juneau, former CRTC chair, former president of the CBC, and briefly, minister of communications in Pierre Trudeau's Liberal government, to lead a group that included former CBC vice-president Peter Herrndorf to review the mandates of the CBC, the National Film Board, and Telefilm Canada. *Making Our Voices Heard: Canadian Broadcasting and Film for the 21st Century* was, according to *Maclean's*, "a radical document, proposing sweeping reform, particularly to CBC television, which it would make virtually commercial-free and almost completely Canadian in content." Its proposal, however, for a new tax on telecommunications services to pay for the CBC produced a fierce reaction from private broadcasters and taxpayers' associations. In the end, "the Juneau report's clarion call for cultural nationalism was all but drowned out by the howls of protest."[41]

The ideas for dramatic change from that report would be echoed in almost every subsequent report about the CBC, as would be the criticisms from private broadcasters and the political right wing. The report was particularly critical of CBC television in both French and English; Juneau suggested that they had moved away "from their public service mandate and become too commercial, too preoccupied with ratings and no longer provide enough of an alternative to commercial broadcasting."[42] The report's recommendations quickly died in the face of political critics and lobbying by cable companies and private broadcasters because no one in the Liberal government was prepared to push for such a dramatic financial overhaul; in retrospect, however, it would have made the CBC much stronger and more distinctive. The political problem at the end of the day was that the Juneau report didn't give anything to private broadcasters—which would have been the price for their acceptance.

In 2003, the House of Commons Standing Committee on Canadian Heritage, chaired by Liberal MP Clifford Lincoln, released a 613-page report that called for the Broadcasting Act to be amended "to include new-media services for the purpose of greater clarity, but that new media be made a central part of the CBC's strategic plan."[43] It recommended that the CBC continue its efforts to move into digital services and that "the future of communications, both in Canada and throughout the world, will be dependent on cross-platform strategies in which online content is used to supplement radio and television programming."[44] While CBC management began to edge in that direction, there was no political leadership for change from the Liberal government. The result was a second missed opportunity to prepare for the industry upheaval that was then starting to build.

David Taras, one of the authors of this book, served as an expert adviser to the committee. He remembers that one of the ideas floated within the committee was for the creation of a "green space" in the cable universe. The green space would be occupied by the CBC and other educational and public service broadcasters and would be free to consumers. All the commercial channels would be available on a "pick-and-pay" basis. The goal was to give the CBC a right of way and a vital lifeline. Nothing came of the idea because, apparently, the CBC turned the proposal down. Taras thought at the time that the CBC's propensity for miscalculation and self-harm was astonishing.

A poignant moment occurred when Taras and Marc Raboy, another adviser to the committee, had lunch with Clifford Lincoln and Charles Dalfen, the chair of the CRTC. When asked about Internet broadcasting, which was still in its infancy, Dalfen admitted that while it was a form of broadcasting, the CRTC had famously decided that it wasn't going to regulate the Internet. The integration of broadcasting with advanced computing and the onslaught of new platforms and services that were in the process of emerging were, therefore, beyond its jurisdiction. The CRTC had relegated itself to the sidelines.

Four years later, in 2007, the same committee held another round of hearings into the CBC's digital and new media activities and how they fit with its mandate under the Broadcasting Act. That report, entitled *CBC/Radio-Canada: Defining Distinctiveness in the Changing Media Landscape*, issued in 2008, had more than 50 recommendations. They included increased, stable, multi-year funding for the CBC; making digital media and emerging technologies an explicit part of the corporation's mandate; placing a greater focus on the Internet; reducing the role of advertising in television; and negotiating a seven-year memorandum of understanding between the broadcaster and government.[45] While the report articulated a way forward that might have changed the CBC's entire trajectory by putting it at the cutting edge of new media development, the Conservative government, like its Liberal predecessor, wasn't interested in changing anything.

The CRTC stayed out of much of this debate. That is not surprising considering that, throughout its history, the regulator in its decisions has never demonstrated any enthusiasm for either the concept of public broadcasting or the CBC, regularly rejecting its proposals and taking no initiative to highlight the role that it perceived public broadcasting played in Canada's media environment. As we commented earlier, the regulator has considered its primary role to be defending and supporting the overall (meaning the privately owned) broadcasting industry. Consumer interests have only recently come to the fore in the face of public resistance to the constantly rising costs of everything regulated by the CRTC, from cable and satellite television packages to mobile phone contracts and charges.

Beginning in 2014, the CRTC conducted a special set of hearings that it called *Let's Talk TV*, which sought to gauge opinions from consumers on the changing television landscape. This was part of a new consumer-centred focus initiated by the CRTC's chair at the time, Jean-Pierre Blais. The process led the CRTC to give consumers new rights in dealing with cable and satellite companies and with technical changes, including redefining Canadian content requirements. But strangely, *Let's Talk TV* didn't include talking about the CBC. The special hearings produced nothing that dealt specifically with public broadcasting and no attempt to probe attitudes on the role that public broadcasting should play in the future.

While the CRTC was ignoring the CBC, the Senate was at work. After a set of hearings about the future of the media, a report from its Transport and Communications Committee in 2006 made a series of recommendations to restrict mergers in the media, but also specifically addressed the CBC. It called for a process to refine the CBC's mandate, followed by a government commitment to long-term, stable financing and a ten-year licence renewal. Part of that restructuring would include an end to advertising and a proposed commitment by the CBC not to duplicate the programming offered by private broadcasters. This included a specific commitment to relinquish coverage of professional sports and the Olympics. The committee also proposed that CBC board members should have a background in programming or journalism, that a parliamentary committee should review appointments, and that the president should be appointed by the government from a list prepared by the board.[46] All seemed common-sense recommendations.

Almost a decade later, the same Senate committee, now stacked with a Conservative majority, took another run at the CBC's future with a new report, *Time for Change: The CBC/Radio-Canada in the Twenty-First Century*.[47] It also recommended that the Broadcasting Act "be modernized to reflect the current environment," spent an inordinate amount of time on recommendations to limit the salaries and external activities of CBC employees, and proposed shutting down all production beyond news and current affairs. The committee also proposed that the CBC should broadcast high-quality programs that were unlikely to be aired by private broadcasters. In other words, the committee wanted the CBC to stay clear of areas that were within the orbit of private broadcasters— which, at the time, seemed to include almost everything on the schedule. It also wanted the corporation to "explore alternative funding models and additional ways to generate revenue to minimize the Corporation's dependence on government appropriations."[48] Liberal Senator Art Eggleton, in a dissenting report, called for an increase in the CBC's parliamentary appropriation to $40 per capita from what was then $29 and the introduction of a new tax on revenues from telecom companies, which would go to the CBC.[49]

Toronto Star columnist Carol Goar dismissed the Senate report as "a shoddy piece of work; poorly researched, internally contradictory, short on vision." She added that nowhere did the report acknowledge that, under the Conservatives, CBC funding had fallen by 18 per cent, adding that the senators "offered no useful advice to the CBC about carving out a niche in today's digital multimedia universe."[50]

The same paralysis seemed to affect the House of Commons Standing Committee on Canadian Heritage when it looked at the state of local media in 2016–17. This time with a Liberal majority, the committee only tangentially touched upon the CBC's activities in the broader context of the problems facing news organizations as they tried to understand and respond to digital shock. Although most of its June 2017 report, *Disruption: Change and Churning in Canada's Media Landscape*, dealt with the decline of newspapers, it did recommend that "CBC/ Radio-Canada prioritize the production and dissemination of locally reflective news and programming by expanding its local and regional coverage, including unserved areas across all of its platforms." Without suggesting new funding mechanisms, this proposal was the equivalent of telling the CBC to build skyscrapers out of thin air. The committee also proposed that CBC/Radio-Canada step away from advertising on its online news platforms.[51] That latter suggestion was in response to complaints from newspapers, which objected to having to compete for online advertising with the government-supported CBC.

In those and others of its 20 recommendations, the Commons committee largely mimicked the proposals from a January 2017 report commissioned by the Department of Canadian Heritage from the Public Policy Forum (PPF). Like the committee's proposals, the PPF report, *The Shattered Mirror*, dealt mostly with newspapers and print-media organizations, but it did make three recommendations about public broadcasting. It called on the federal government to ensure that the CBC paid more attention to the section of its mandate that stated that informing Canadians was a primary role for the broadcaster, and it proposed an expansion of what it called "civic-function" news.

The report also recommended that the "CBC should move to a system of publishing its news content under a Creative Commons licence, marking the next logical step of a public-service news supplier in the digital age." It insisted that "such an open-source approach would go a long way toward moving the organization from a self-contained, public-broadcasting competitor to a universal public provider of quality journalism. It would strengthen the media ecosystem overall, anchoring it in greater integrity and maximizing the reach of CBC journalism."[52] While sounding revolutionary, the fact that no successful news organization has tried to go open source and that the costs of policing and administering this news model could be extraordinary should give some pause.

Finally, the PPF called for the CBC to phase out online and other digital advertising within a year. Unlike the House of Commons committee's report, the PPF suggested that the corporation's parliamentary appropriation increase by the amount the CBC would lose in giving up all digital advertising on its online sites. Translation: this would not mean much of an increase.

The CBC responded quickly to the PPF, enthusiastically endorsing its role in informing Canadians, but had little time for the idea that its news content should be open-sourced with others. In a late January 2017 news release, the corporation said that the concept "needs further study, particularly given the growing concern over the accuracy and behaviour of some online sites. Canadians need to know they can trust the integrity of CBC/Radio-Canada journalism."[53] It also characterized the proposal to end digital advertising as a half-measure, stating:

> In our proposal [to the committee] we recommended the government develop a cohesive cultural investment strategy and that CBC/Radio-Canada remove advertising from ALL of its platforms.... Removing only digital advertising would mean CBC/Radio-Canada would still be a competitor for advertising revenue on other platforms which would limit its ability to partner with other organizations. Also, because modern advertising is sold in bundles including both digital and television, partially removing ads would hurt overall revenue.[54]

What was perhaps most remarkable about *The Shattered Mirror* was how long it seemed to take the "powers that be" that were behind the report to realize the extent of the tsunami that was engulfing them. The reality is that the digital storm has long since made landfall and that, despite years of experimentation and struggle, many in the Canadian media elite were far too late in seeing it coming. The weakening and impending collapse of Canada's media institutions also seems to have taken governments, regulators, and MPs by surprise. One of the most telling findings in the report was that because members of the public had so many media choices at their disposal, at least half of Canadians failed to realize that their own media system was collapsing around them.

The Failure to Act

By 2019, it all added up to a wasted quarter-century that could have prepared public broadcasting for the new attention economy. While there was much wilful blindness, there were lots of creative ideas about how the federal government should respond. But nothing happened.

Politicians and federal governments consistently and consciously chose to do nothing despite being told what needed to be done every time a committee held a new round of hearings. At the very least, the Broadcasting Act could have been updated to redefine the CBC's role in a digital world, a new funding regime that provided the CBC with long-term, stable funding could have been put in place, advertising could have been eliminated, and a credible process for appointing a knowledgeable board of directors and president could have been implemented. These small but reasonable steps were not taken.

The federal government consistently missed opportunities, lacking imagination and any stomach for risk-taking or innovation. Politicians all seemed much more fearful of how public broadcasting could damage their short-term political interests and future than intrigued by what it could do in a globalizing world to portray an image of the diversity and complexity of the country to its own people as well as to others around the globe.

Regulators missed opportunities by spending much of the period implementing policies designed to enhance and protect the interests of private broadcasters, specialty-channel owners, and cable or satellite companies, to the detriment of the CBC. The CRTC consistently failed to see the public broadcaster as having any broader role beyond simply being another player at the table. It regularly refused the CBC's requests for specialty-channel licences, which could have contributed to giving it a more secure financial base, thereby allowing it to move beyond CBC News Network to broaden its revenue sources to include a significant amount of direct subscriber income from viewers—the tonic that has kept private broadcasters alive as advertising disappears. Even the CRTC's new-found consumer focus missed an opportunity to lay out a vision for public broadcasting.

The CBC's board and management were also guilty of missing opportunities. The political appointees on the board of directors were too willing to put the interests of the people who appointed them ahead of the interests of the organization they were overseeing. CBC management's response to the slashing of its parliamentary appropriation, which occurred with the same regularity as parliamentary reviews of the CBC, also lacked courage. Managers took the easy way out by avoiding tough decisions about eliminating specific activities. In the process, they weakened everything the CBC did, whether on radio or television. There was no move to give up some activities in the hope of concentrating the remaining resources so that the CBC could be better in those areas than anyone else—no move to carry out a strategy to build audiences and guarantee survival.

It was a quarter-century of contraction of ambition, innovation, international scope, breadth of activity, and ultimately, audiences as Canadians noticed the changes and reacted negatively. The CBC had shrunk to the point where it was a shadow of what it once was. Most significantly, though, there was little attempt to try to get ahead of the steamroller of the attention economy, which would undermine the entire broadcasting system and, with it, the rationale for a public broadcaster.

Chapter 4

The CBC in the Digital Storm

The essential problem for the modern media environment, writes James Webster of Northwestern University in Evanston, Illinois, is the "widening gap between limitless media and limited attention." Founded in a time of media scarcity, when it was the only game in town in many communities and where, even in large cities, it faced only a handful of rivals, the CBC now finds itself having to compete in an era of hyperabundance. In programming areas that are central to the CBC's mandate, such as news, drama, sports, and children's programming, it faces a dizzying amount of competition. While the CBC is still a main showcase for Canadian content and faces surprisingly little competition in this realm, finding and holding its audience has never been more difficult. In an era of continuous partial attention, delivery platforms that use algorithms to contour the news to individual characteristics and interests and a streaming revolution that is growing almost exponentially, surviving the digital storm will not be easy.

This chapter will discuss three challenges that the public broadcaster faces in the new attention economy. We will first examine the sheer amount of competition that the CBC must contend with, wherever it turns. Its problem is that it has to compete against companies that offer cable, Internet, and cellphone services in addition to being publishers and broadcasters, while it remains confined to broadcasting and digital platforms. Wherever it turns, the CBC faces multimedia platforms that enjoy economies of scale and access to capital that are well beyond its reach and capacity. We will discuss both the major American giants as well as the nature of the CBC's Canadian competition.

Second, the chapter will describe how Netflix, in particular, has altered the basic geometry of Canadian TV. While the old broadcasting models now exist side by side with the new, the size of this new media wave gets larger and larger every year. As mentioned earlier, the number of viewers under 30 who are cord-cutters, or cord-nevers, is approaching 50 per cent. In cutting themselves off from the traditional broadcasting system, they are also saying goodbye to much of the Canadian content programming that it delivers. They also diminish the capacity of satellite and cable providers to contribute to the Canada Media Fund, which supports Canadian programming.

Last, we will describe the particular challenges brought by Facebook. The social network has become one of the principal means by which the CBC distributes material posted on the CBC and individual-program Facebook pages. It acts as a conveyer belt, delivering stories, programming notes, and highlights to users. Facebook, which, in some ways, is the new public space, is critical to the CBC in at least two ways. The recent decision to downplay news and give priority to exchanges between friends can only diminish the CBC's presence within Facebook. In addition, Facebook threatens the CBC and other media organizations with "brand annihilation" as its stories are lost in an endless sea of other Facebook stories. More critically, Facebook poses important questions about the future of public spaces and democracy—the very questions that are critical to the CBC's mandate.

There are at least two points to make about the new attention economy. First, the sheer volume of what is available to users is staggering. The daily smorgasbord for media users is simply colossal. On any given day, there are more than 3.5 billion Google searches; more than one billion hours of videos are watched on YouTube; an average of 1.3 billion people log onto Facebook, with users posting close to 3 billion "likes"; 40 million photos are posted on Instagram alone; and there are over 500 million tweets.[1] ESPN sends out more than 700 million alerts per week to mobile phone users. This is aside from the more than 1 billion websites strewn like leaves across the length and breadth of cyberspace. Given the proliferation of TV services, on cable and satellite and through streaming platforms such as YouTube, that effectively operate TV channels, the 1,000+ TV channel universe has finally arrived.

When it comes to music, Pandora alone plays 21 billion hours of music a year. Spotify has a song bank of 30 million tracks, and more than 5 billion tracks have been streamed on its Discover Weekly playlists since 2016. Add to this Apple Music, Amazon Music Unlimited, Pandora, Tencent, SoundCloud, Google Play, Deezer, Rdio, Tidal, and Last.fm, among other platforms, and the choices are almost infinite. On Spotify, choices can be made not only from the old, predictable genres but also from mood categories such as "wonky," "downtempo," and "stomp and holler." IBM's Watson can digest 8 million pages a second, conduct an almost instant analysis of virtually any issue, and zero in on any single news item in real time—most of which is accessible to the public. Then there's Netflix, which is rapidly turning into a death star for Canadian broadcasters. A survey conducted in 2017 found that 53 per cent of Canadians subscribed to a streaming service, and Netflix, with close to 7 million subscribers, was far ahead of the other 20 subscription services available in Canada.[2] Simply put, the media sky now resembles the Milky Way, and the CBC's star, however glittering it might be, is sometimes hard to find, even with its introduction of Gem, its own streaming service.

Moreover, much of media production is bottom up. The great majority of people are media producers in their own right. Derek Thompson of the *Atlantic* once observed, "Every social media account, every blogger, every website, and every promiscuously shared video is essentially a radio station."[3] But it's even more than that: users curate content, make and redact videos, post photographs, act as citizen journalists, and comment on and often bear witness to events. Moreover, as Michael Smith and Rahul Telang point out, improved technologies such as the Canon 5D Mark III camera with a Carl Zeiss high-powered lens and the Final Cut Pro editing suite have allowed individuals to create Hollywood-grade media products at relatively little cost. On another level, would-be producers now have access to YouTube's production facilities located around the world, with YouTube in some cases acting as an angel investor. In addition, start-ups have never been easier to launch. New funding sources such as Kickstarter and RocketHub now provide millions of dollars of support for TV programs and movies.

What's remarkable in this vast media scramble is that some artists and intellectuals have become Internet stars in their own right, with larger followings than many conventional media organizations. YouTube megastar PewDiePie attracted over 83 million subscribers in 2019, and video makers and performers such as Ed Sheeran, Dude Perfect, and Konrad Cunha Dantas have over or close to 40 million subscribers. YouTube's *Epic Rap Battles of History* has close to 15 million subscribers and has generated billions of views. Chinese dissident and global icon Ai Wei Wei has close to 400,000 Instagram followers, a fact that has not only kept the Chinese government from putting him in prison but also galvanized a global audience for his installation art, films, and architecture. In Canada, music superstar Drake has 37 million Twitter and more than 40 million Instagram followers, and his YouTube videos typically attract more than 200 million views. YouTube sensation comedian Lilly Singh, from Scarborough, Ontario, whose signature is IISuperwomenII, has 14 million subscribers and is now the host of the late-night NBC talk show *Last Call*. Canadian poet Rupi Kaur has 1.6 million Instagram followers, a Twitter following of close to 200,000, and her books *Milk and Honey* and *The Sun and Her Flower* have sold millions of copies around the world. Intellectuals such as philosopher Jordan Peterson have used the Internet to place themselves at the centre of public debate and interaction. An estimated 40 million people, for example, have watched his videos about morality, discipline, and manhood, which helped catapult his book *12 Rules for Life* onto the bestseller lists in 2018. Amazingly, Poppy, a social media fembot, has some 50 million followers, and her YouTube videos have been viewed close to 300 million times.

In each case, these new media superstars have used multimedia platforms to propel themselves into the global marketplace. Needless to say, the CBC would be pleased if any of its stars or shows could attract a similar following. But what is interesting is that, until the 1990s, arguably no Canadian artist or public figure could become popular without access to the CBC. It was the CBC that helped make them stars. Today, an artist such as Drake can have a national and global following without any help from the CBC.

Media abundance may be the biggest threat that the CBC faces. Not only is the public broadcaster challenged in virtually every program category that it once dominated—news, sports, comedy, children's programming, musical variety, etc.—by an avalanche of competition but also, in some areas, competing may no longer be possible or even logical. For instance, according to Derek Thompson, in 2000 there were roughly 425 scripted series and reality shows broadcast annually by conventional US networks and cable franchises.[4] By 2018, the number of scripted, original streaming, cable, and broadcast series—excluding reality shows—had surpassed 500.[5] The number was expected to increase even more as Disney Plus and Apple TV+ as well as increased investments by Netflix and Amazon Studios come to air in 2020. As Thompson points out, in this new age of *peak TV*, audience sizes for virtually all shows have shrunk dramatically, and the chances of a single show becoming a breakaway hit are about the same as winning the lottery.[6] In the past, scoring a series of hit shows would create a financial bonanza for conventional broadcasters that would keep them humming for years. But today, the risks seem to outweigh the rewards. For Canadian broadcasters, the economics are stark. Why risk investing large sums on shows when the chances of success are so small? Better to avoid risk, take fewer chances, and hope for the best.

According to Harvard business professor Anita Elberse, however, the strategy of placing a large number of small bets is far riskier than making a small number of large investments. As she explains, "The highest performing companies in the entertainment and media sector thrive by investing a relatively large proportion of their resources in just a few titles and then turning those choices into successes by giving them a higher level of development and marketing support. It may be partly a self-fulfilling prophecy, but it works."[7] Major blockbuster events such as the latest Marvel movie, album release, or awards show can take up a lot of space in the culture-dominating media and online traffic for weeks at a time, sidelining or blanketing out other events. The CBC's programming philosophy, which is the product of having to produce a relatively large number of programs with a limited budget, is the exact opposite of the strategy used by a number of other industry players, who bet, often with the benefit of audience analytics (analysis about their audiences and their online viewing habits), on what they hope will be blockbusters. So the strategy for many industry players is to swing for the fences, even if this means more than a few strikeouts.

Thompson summarized the problem faced by both the CBC and other media companies perhaps best: "In this bottom-up world, where cultural authority shatters into a million channels of exposure, the hits are harder to foresee—and authority is harder to protect."[8]

Marketing expert Sam Alter describes the attention economy as being similar to a budget in which time is limited and there is space for only a very few choices.[9] There is also another, more stringent budget process at work. At a time when many Canadians are just a paycheque away from falling off a financial cliff and can afford to spend on only so many subscriptions and micropayments, the race for the audience's attention is all the more crucial. This cruel winnowing process allows only a small number of companies to survive. The CBC's great advantage in this Darwinian world is that its main channel and online presence are freely available to audiences, including its Gem service—although there is a charge for Gem Premium. Its great disadvantage is that, as we discussed in an earlier chapter, virtually every other media company now has a stake in bringing it down.

One of the most interesting studies of the corrosive effects that media abundance has had on democracy is by Princeton political scientist Markus Prior.[10] His analysis was based on co-relating TV audience numbers in the United States with voter turnout in elections in that country as TV went from a relative scarcity of channels in the 1960s and 1970s to the surplus brought by the explosion of cable and satellite channels in the 1980s and 1990s. His central finding was that once audiences were given access to a smorgasbord of entertainment choices, audiences for news and public affairs plummeted, and as a result, so did voting. Whereas viewers could not avoid at least some exposure to news and politics when there were only three or four channels to choose from, with the proliferation of channels they could avoid news and politics almost entirely. To put it more dramatically, the more entertainment that is available, the more likely people are to watch only entertainment, crowding out news stories. Derek Thompson notes that the typical viewer watches only five minutes of cable news a day.[11] Matthew Hindman calculates that only 5 per cent of time on mobile devices is spent watching or reading hard and soft news.[12]

Interestingly, Robert Putnam spun a similar hypothesis about entertainment crowding out news in his controversial book, *Bowling Alone*, published in 2000. He argues that the erosion of community involvement and social capital that has taken place over the last several decades has been caused mainly by the spiralling rise of entertainment TV. As Putnam observed, "Nothing—not low education, not full-time work, not long commutes in urban agglomerations, not poverty or financial distress—is more broadly associated with civic disengagement and social disconnection than is dependence on television for entertainment."[13]

While some might argue that the entertainment thesis is overdrawn and out of date, and that there is a renewed interest in news and in voting by younger Canadians, an entire cottage industry of scholars and observers has emerged who believe that the lessons of democracy are being lost on younger citizens. Harvard professors Steven Levitsky and Daniel Ziblatt, for instance, argue that "the guardrails of democracy" have collapsed, in part because the old media system has mutated.[14]

A second point about the new attention economy is that each user can now construct their own customized media bubble. Users create their own media worlds out of a vast quilt of sources: Facebook, YouTube, Reddit, Twitter, Netflix, iTunes, newspapers, Instagram, etc. And while there is much crossover among media worlds, no two media environments are the same. While during its so-called golden era, from the 1930s to the 1980s, the CBC arguably constituted the main public sphere in Canada, the principal meeting place where Canadians gathered to witness and experience events, the notion of a single public square such as that envisioned by Jürgen Habermas has vanished forever.[15] While mass audiences can form around certain issues and events, they are ever shifting and reconfiguring. The audience is fleeting and liquid.

Michael Wolff points out that even the simple act of measuring has become elusive and almost impossible. As he observes, the problem is "not the people gathered to pay attention. But people moving to and fro, taking a more often than not random path, seeing little, absorbing less (attention, that is, time on page, measured in fractions of a second). The better metaphor surely was a highway billboard slipping by at sixty miles an hour than a thirty-second spot."[16] Moreover, according to Wolff, nothing can be taken for granted: "Every eyeball has to be captured again—as though for the first time."[17] The audience is constantly up for grabs. While the CBC still has a loyal audience, particularly in radio—better educated, older, more nationalistic, and more rural—at any given moment it can gravitate elsewhere or disappear.

Compounding the situation is that there is now a growing literature about users living in a state of continuous partial attention.[18] Addicted to multi-tasking, needing endorphin hits, attending to smart phones every few minutes, and frenetically switching channels when they become bored, they pose a difficulty for producers: not only finding listeners and viewers but also keeping them listening and watching for an extended period of time. In the age of constant distraction, media products have to have a "sticky" quality, have to be able to keep audiences absorbed for long periods, in order to be successful.

Florian Sauvageau, a former Radio-Canada host and professor emeritus at Université Laval, has gone so far as to argue that viewers now want "snacks" rather than "meals."[19] Their nervous systems are now geared to texts, short videos, tweets, apps, scrawls, vines, highlight packages, 10-to-15-second ads, alerts, and Snapchat stories rather than to reading long articles or books or sitting through anything that lacks the quick jolts that they have become accustomed to or the endorphin hits that their brain finds chemically rewarding. TV producers worry that many young viewers no longer have the attention spans needed to watch an entire hockey or football game or hour-long show. In fact, Chartbeat, which measures, in real time, online audiences' minute-by-minute engagement with individual stories, found that less than 10 per cent of readers actually scroll to the end of an article. Most don't reach the mid-way point.[20]

Sauvageau's point is that while snacks might be light and sugary, they lack the nutrition provided by healthy media meals. Arguably, the entire purpose of public broadcasting is to provide the in-depth stories and analysis that so many others are failing to provide. The danger, of course, is that the very act of providing those longer background stories can be a death knell in a digital world where the audience has become used to digesting information in bite-sized pieces.

Of course, people binge-watch TV programs, play video games for hours on end, and read long novels, but the architecture of storylines has to be captivating and sticky. Audience expectations of production quality are arguably so much higher today than they once were.

Media executives once believed that the way to survive the onslaught was to remain above the chaos as a mediator and organizer. Conventional news organizations would decide what information and sources their readers or viewers would pay attention to. This turned out to be little more than a pipe dream as search engines and platforms such as Google, Facebook, Amazon, and Netflix were able to rank and optimize choices on a much grander scale. It wasn't long before conventional news organizations were the ones being ranked rather than the ones that ranked others. Some media platforms such as the *New York Times*, the *Guardian*, the BBC, Reddit, and the *Washington Post*, to name a few, had the vision, dexterity, and financial resources needed to rise above the fold and become the new general-interest intermediaries. Most others, and as we argued in the introductory chapters, the CBC among them, were pushed out of this vital role.

This is especially the case with recommender systems. As Hindman reports, sites and apps that have recommender systems are far stickier and more successful than ones that don't have them.[21] Without the capacity to keep data on its users, the CBC can't be a recommender site in the same way that its rivals are. It simply can't direct traffic for its users. This is a point that we will discuss at greater length later in the chapter.

At the very least, the CBC's place as a chief presenter and curator of Canadian programming and culture has become more difficult to maintain simply because audience members have also become curators and distributors. To some degree, the power equation has been upended. The CBC now has to depend, at least to some extent, on its audience to be its transmission line rather than the reverse.

Winners Take All

While the Internet provides users with a vast kaleidoscope of choices, it is also a merciless, winner-take-all environment. There are a small number of winners—mammoth whales such as Facebook, PayPal, eBay, iTunes, Netflix, Instagram, Google, Amazon, YouTube, Disney, Alibaba, Tencent, and Twitter, among others—and millions of minnows. The minnows get devoured, are soon forgotten, or survive on the edges. Facebook, Apple, Amazon, Netflix, and Google (colloquially known as FAANGs) have a collective wealth that exceeds all but six of the world's economies. While barriers to entry are low, so there is an illusion that any site can grow and become a major player, the reality is very different. The giants benefit from what economists call "network effects": the more they are used, the more data they collect and can utilize, the more important they become, the more time people spend on them, etc. It is similar to a "constantly compounding rate of interest," with wealthy sites continually getting larger and more powerful.[22] The vast majority of web traffic goes to a relatively small number of sites. In fact, the four largest sites account for one-third of all web traffic.[23]

The same can be said for news sites. The online public gravitates to a small number of news powerhouses such as the *New York Times*, the *Guardian*, *Google News*, the *Wall Street Journal*, and *BuzzFeed*, among others. Media guru Robert McChesney believes that there is little in the way of a middle class on the web.[24] Most other sites exist as part of a long, thin tail. They collect niche audiences, appeal to scattered groups with particular and narrow interests, and rarely become anything more. The CBC is still arguably a member of the middle class—at least for now.

According to James Webster, one of the iron laws of media economics is that the "more abundant the medium the more concentrated audiences tend to be."[25] The biggest winners are the FAANGs. While each of them has, to use José Van Dijck's description, its own "culture of connectivity," what they have in common is that they are gateways and toll booths through which other media has to pass to reach its audience.[26] To put it differently, these platforms monetize the music, videos, ads, and performances that they host, often leaving performers and producers with very little remuneration. The key to the kingdom for these platforms is that they not only control these vital bridges but also take a sizable cut from media producers for granting them access to the public.

The FAANGs are also voracious collectors of data—data that is provided willingly and at no cost to them by their users in exchange for free use of their services. Sophisticated algorithms target the needs and preferences of their customers, including their locations and travel patterns, likes and dislikes, the words and phrases that are likely to be the most convincing, people's price thresholds, and buying histories as well as the likes and dislikes and the buying habits of their friends. These companies know more about their users than their mothers or lovers ever will. While policies differ among companies, data is used to match advertisers with consumers and, in some instances, is sold to third parties or, in the case of Facebook, given to "partner" businesses and organizations.

It's important to note that while the CBC keeps statistics on the nature and size of its audiences, it is limited in the amount and quality of data that it can collect. Simply put, a Crown corporation is not in the business of creating individual digital profiles based on knowing each citizen's consumer preferences, favorite keywords, friends and contacts, and previous online visits. Keeping such intrusive personal information would be a major scandal. But this is not the case for the CBC's rivals.

They are free to use data analytics to plan programs and schedules, provide advertisers with instant feedback, and target their users with customized stories or messages. The CBC, and public broadcasters in general, are almost defenceless against platforms for which data is both a main product and their principal weapon. As media becomes more and more about collecting and using data to chart audience behaviour and interests and then using that data to make programming decisions, the CBC risks not only losing the game but not even being on the field.

Most crucially, these giant industrial platforms are also media producers and commissioners of programming themselves. So-called tech media comes in many forms. Google owns YouTube and YouTube TV. Facebook, for instance, owns Instagram and WhatsApp; hosts Facebook Live and Sound Collection; has the broadcast rights for the Union of European Football Associations Champions League, major league soccer, and a batch of Major League Baseball games; and makes news decisions for billions of people every week. Amazon's founder and chief executive officer (CEO), Jeff Bezos, owns the legendary *Washington Post*, which together with the *New York Times* has led much of the journalistic crusade against the Trump administration. In addition, Amazon Studios spends billions of dollars every year on TV and film production; the company owns Twitch.tv, with access to its large, younger audience; Amazon Prime has the rights to broadcast NFL games on Thursday nights; and Amazon Web Services is the world's largest cloud company, counting among its clients a myriad of US intelligence agencies, including the Central Intelligence Agency.

While Netflix pays for the right to showcase TV programs and movies made by major producers from across the globe, it's a major producer in its own right. In fact, Netflix spent US$8 billion on original productions in 2018, more than six times the CBC's entire budget and more than 25 times its budget for English-language TV. Netflix will spend at least US$10 billion in 2019. Its original series include *The Crown*, *Stranger Things*, *House of Cards*, *Orange Is the New Black*, and *Ozark*, among hundreds of other shows. Apple has joined the fray and is now a movie and TV producer, with its own Apple TV+ streaming service.

Nor is the influence of the tech giants limited to hosting other media and producing their own media products. Their tightening economic grip is evident in almost every sector of the economy. Google makes robots, develops autonomous vehicles, and is a major player in artificial intelligence and in developing blockchain technology. It also owns Nest, a company that specializes in creating the "wired home," including home protection and saving energy, and Waze, which helps drivers navigate through traffic. Google is also a major investor in Airbnb and Uber.

Amazon has built the world's most efficient logistics and transportation system; it even has its own trans-Pacific shipping company so that it can take advantage of trade with China, India, and Japan. It owns Amazon Robotics, which manufactures warehouse robotic fulfillment systems; is developing its own fleet of drones; purchased the Whole Foods grocery chain; and is experimenting with retail grocery and book-and-electronics stores. Jeff Bezos owns Blue Origin, a company that launches satellites into orbit and has taken the first steps in space tourism. Facebook also manufactures drones but is also a leader in sound production and in creating virtual environments. Apple has a solar energy division, has an interest in autonomous vehicles, manufactures mobile phones, and is a leading software producer.

For the CBC and other media organizations, the FAANGs are both close allies and strong competitors. On the one hand, they are hosts and access points that are crucial for reaching audiences. With these media platforms accounting for every one of the ten most visited sites, there is no way to go around them, sidestep their influence, or reach over the castle walls that they have created. For instance, many Canadians learn about what's on the CBC or access it through Facebook, YouTube, or Apple products. At the same time, the shows, videos, movies, and TV programs produced by these same tech media platforms compete directly against the very companies, such as the CBC, whose products they host or advertise.

Tech media has changed the basic geology of the media world. Because these companies' activities are largely unregulated by the US and Canadian governments, they have been allowed to accumulate extraordinary and unbridled power. The scale and sophistication of their data collection, their dominance over advertising, and their monopolistic and predatory practices have entirely redefined the nature of the media marketplace. This regulatory free-for-all is not as true in Europe, however. The European commissioner for competition, Margrethe Vestager, fined Google 2.42 billion euros in 2017 for giving preferential treatment to its own products, including forcing competitors to use Google as the default search engine on smart phones, and imposed a 13 billion-euro fine on Apple for working with the Irish government to avoid taxes. In 2019, the European Union (EU) again fined Google, this time for US$1.7 billion, for blocking ads by non-Google companies on third-party sites. Facebook, Twitter, Microsoft, and Google have all signed on to an EU code of conduct that obligates them to take down hate speech within 24 hours of its appearance, as well as publish a corrective article, so that bigotry does not go uncorrected or unchallenged.

In 2019, the Federal Trade Commission in the United States levelled a $5 billion fine against Facebook for its data breaches and opened up investigations into Google and Amazon as well. Investors considered the fine to be so small compared to Facebook's revenue that the stock price was barely affected.

Unlike their European counterparts, Canadian regulators have largely sat on their hands and done little. The general feeling is that these new platforms are beyond the control of the Canadian state and that the Internet can't and shouldn't be regulated. While the federal government can negotiate deals such as we saw the Trudeau government do with Netflix in 2017, it has been reluctant to interfere in the marketplace even when that marketplace no longer reflects the national interest or jeopardizes the very existence of Canadian industries.

Another harsh reality is that for all the democratization of media production, most of it is now concentrated in a very few hands. Corporations such as Disney, Time Warner, Comcast, Viacom, and Sony dominate the media landscape in ways that were inconceivable just a few years ago. After its takeover of 21st Century Fox film and TV studios in 2019, Disney's revenue topped US$90 billion. To place this in perspective, Disney's revenue is more than 90 times the CBC's annual allotment from Parliament. Disney's stable of brands includes ABC, Pixar Animation Studios, Lucasfilm and its Star Wars franchise, Marvel Entertainment, Sky TV, Fox studios, and sports channel ESPN. This is in addition to a smorgasbord of theme parks; hotels; stores; cruise ships; Broadway and Las Vegas shows; radio and TV stations; cable franchises such as the Disney Channel, A&E, and FYI; and streaming services, including Hulu, Disney+, and ESPN+. Disney also owns BAMTech—a state-of-the-art data and graphic design company.

Disney is the worldwide leader in the sale of licensed merchandise, and its annual revenue from selling consumer products alone would keep the CBC running at current levels for at least half a decade.

As dominant as Disney is on a variety of platforms, Comcast is substantially larger and arguably even more powerful. In addition to being the king of cable and a leading Internet service provider, it owns NBCUniversal, Telemundo, DreamWorks Animation, Xfinity, and MGM, among a host of other media treasures. With revenue in 2018 of more than US$95 billion, Comcast is larger than the entire media industry in Canada. AT&T-Time Warner is bigger still. With revenues approaching US$150 billion, it brings in the equivalent of the budget of Ontario every year.

What is interesting is not only the sheer size of these super-conglomerates but also the degree to which they often cooperate to limit competition. While fiercely competing against each other in some areas, they have created strategic alliances in other areas. They cooperate, for instance, in scheduling movie releases and sports events so that their media products don't compete directly against each other. TV studios, for instance, won't bid against a rival studio when TV programs come up for renewal. They join forces to lobby governments and often work together on joint ventures. Columbia University law professor Tim Wu has compared these corporations to "a gang of octopuses."[27]

Where once a handful of companies would dominate a single industry, so that music labels produced music, movie studios made movies, newspaper companies produced newspapers, etc., digitization allowed companies to expand far beyond the confines of a single sector. Media organizations realized that unless they moved quickly to occupy new territory, others would be there before them. For these major conglomerates, the choice was simple and brutal: either expand or die a slow death. Moreover, size mattered. Companies such as Disney and Comcast could invest large sums in takeovers, absorb losses that smaller companies couldn't survive, and heavily promote their products and stars.

What has happened to the Canadian film industry is a stark example of how an entire national industry can be sidelined by these major conglomerates, even within the boundaries of its own country. Because Hollywood studios control distribution, few Canadian films make it into theatres. Even the best films are relegated to a second tier of film festivals, pop-up film clubs, last-chance theatres, and late-night or early-morning cable, where few Canadians will ever see them. The CBC can provide another fleeting window. Before long, even the best Canadian films simply disappear from sight.

The CBC faces a similar alignment of forces in Canada, although on a smaller scale. While it is confined to broadcasting and online publishing, all its main rivals are much larger multimedia conglomerates. Bell Media, for instance, which is the majority owner of the CTV and CTV 2 networks, specialty channels TSN 1–5, MTV 1–2, RDS 1–2, BNN Bloomberg, CP24, the CTV News Channel, as well as Discovery Channel and The Comedy Channel, among a myriad of other media properties, is part of Bell Canada Enterprises (BCE), the giant mobile phone and Internet service provider. Together with Rogers and business mogul Larry Tanenbaum, Bell also owns Maple Leaf Sports & Entertainment (MLSE), which includes the Toronto Raptors, the Toronto Maple Leafs,

Toronto Football Club (Toronto FC), the Toronto Argonauts, and the Toronto Marlies in its stable as well as Toronto's Scotiabank Arena (formerly Air Canada Centre), among other properties. It also owns 18 per cent of the fabled Montreal Canadiens hockey team.

BCE's revenues are more than $22 billion—roughly 20 times those of the CBC. Rogers, whose revenues are more than 15 times the CBC's annual budget, is even more diversified. It is a cable, Internet, and wireless provider; has a home-monitoring business; and published a fleet of magazines that included *Maclean's* and *Chatelaine*, among dozens of other titles, until it sold them in 2019. Its broadcast properties include the Citytv network, Omni Television, Sportsnet channels, The Shopping Channel, and FX as well as more than 50 radio stations, including Sportsnet 590 The Fan. As mentioned above, Rogers, together with BCE, owns MLSE, which is at the summit of the Canadian sports universe, and is the sole owner of the Toronto Blue Jays and its stadium, the Rogers Centre. As discussed at greater length in the next chapter, in 2014 Rogers spent $5.2 billion on the rights to broadcast NHL games until the 2025–26 season.

Québecor is Radio-Canada's main rival in Quebec, and it is an inescapable part of the province's culture. Wherever one turns in Quebec, Québecor is there. It is, through Vidéotron, the province's largest cable and Internet provider; owns TVA, the most viewed TV network; and a bevy of key specialty services, including channels for news, music, and sports. The conglomerate also owns *Le Journal de Montreal* and *Le Journal de Québec*, respectively the second- and third-largest-circulation newspapers in the province, as well as *24 Heures*, a tabloid newspaper available for free. It is also a book and magazine publisher and has a live entertainment arm. It bought the French-language rights to broadcast NHL games from Rogers in 2014. If it weren't for Radio-Canada, the Quebec media horizon would be almost completely dominated by Québecor, with its voices often being not only the loudest but also the only ones heard.

A fourth major Canadian competitor is Corus Entertainment. Both Corus and Shaw Communications are publicly traded companies controlled by the Shaw family of Calgary. Shaw Communications is the main cable provider in western Canada and owns Freedom mobile, Canada's fourth-largest mobile phone provider. Corus runs the Global TV and radio networks; has a number of CTV affiliate stations; operates more than two dozen specialty channels, including Food, History, the Disney Channel, National Geographic, and HGTV; as well as radio stations and a satellite TV service. Combined revenues are about the same as the CBC's total revenue.

While none of its rivals can rest easily as conventional TV, cable, and satellite subscribers continue to decline, the CBC is nonetheless at a disadvantage. This is because its competitors enjoy economies of scale, can offer lucrative deals to advertisers, and can promote their programs across platforms. The CBC is not a media company in the same sense that its competitors are. For instance, as discussed above, Anita Elberse argues that one of the ways that media conglomerates win the entertainment "arms race" is by employing a blockbuster strategy. The strategy is devilishly simple: sign up big stars and then create an audience through large promotional budgets and a cross-media campaign. The conglomerates "push" opening weekends, TV series, books, concerts, and sports events to the point where they shape the cultural agenda. As one media executive told Elberse, "You can spend so much that audiences will show up."[28]

While Canadian media companies can't compete against global conglomerates, companies such as Québecor and Rogers can, from time to time, mount major promotional campaigns that can capture public attention. This is especially the case with Québecor in Quebec. The reality is that the CBC is not in the same league as its large US competitors and has few of the advantages of its Canadian competition. While its rivals may not be too big to fail, the question is whether the CBC is now too small to succeed.

How Streaming Changed the Game

Two caveats before we begin. First, while for most Canadians streaming remains a supplementary system, an addition to rather than a replacement for conventional and cable broadcasting, for a large number of those under 30 streaming is their main source for movies, TV shows, and even sports programming. It is their first and last stop—with no stops in between. Second, while the market for streaming services is becoming crowded—with close to 20 such services in Canada alone, Netflix maintains a large lead: over 50 per cent of Canadians had a Netflix subscription in 2018.[29] It has by far the most subscribers, produces the most programming, and benefits from "network effects" so that the more subscribers it has, the more data it collects, and the more money it can spend on original series, which, in turn, produces more subscribers. According to a filing made by Corus Entertainment with the CRTC, in 2017 prime time viewing of Netflix by Canadians aged 25 to 54 exceeded the time that they spent watching CTV, Global, and the CBC.[30] According to one estimate, by 2020 more Canadians will subscribe to streaming services than to cable TV.[31]

In their discussion of how Netflix changed the nature of TV viewing, economists Michael Smith and Rahul Telang stress the fact that Netflix is, among other things, a data collection machine.[32] Every time a subscriber watches a movie or a TV program, their selection becomes part of their digital signature. Netflix knows your name; knows how often, when, and for how long you watch; and knows what you like. Unlike broadcasters, which go through the now antiquated system of choosing among pitches and scripts, then testing a potential new show by making a pilot, which is broadcast and then run through a wringer of audience testing and reactions, Netflix has more of a fine-tuned sense of what its audiences want.

According to Smith and Telang, when Netflix first decided to make *House of Cards*, sinking US$100 million into the first two seasons, or 26 episodes, it did so after other studios had turned it down. While the accepted wisdom among production studios at the time was that political dramas wouldn't succeed, Netflix's data told a much different story. A large number of subscribers liked films and TV shows that had been directed by David Fincher, adored Kevin Spacey (at the time, at least), and had rented DVD copies of the original BBC series. When the series' creators told Netflix that they wanted "to tell a story that would take a long time to tell," Netflix didn't hesitate because the numbers told them that it would be a hit.

Another example of Netflix's straight-to-series formula was *Stranger Things*, which filled what the data suggested was a giant hole in the schedule for a big-budget production aimed at young adults.[33]

Although Netflix does not release ratings and clearly has had flops, it pretty much knows where its original productions will land along the long tail from mass-audience blockbuster to small, cultish following.

Netflix's principal innovation was to release an entire season all at once instead of showing episodes once a week over a period of months. Not only did this give the audience much more control, but it also reshaped the culture of TV viewing. Binge-watching, which is arguably more in line with the way that people want to experience TV, has for many viewers become the new normal. Subscribers will often do a deep dive into a series, devoting entire weekends or evenings to watching episode after episode. Andrew Romano has argued, "Serialized, streaming TV is tailor-made to keep the endorphins flowing," and so viewers are reluctant to tear themselves away from what is immersive and pleasurable.[34]

To keep audiences hooked, Netflix has a 5 to 20 -second countdown between episodes (depending on the show) so that it takes little effort to continue watching. The ball is kept rolling by a recommender system, which suggests shows similar to the one that you have just watched or that Netflix's algorithms suggest you might like. The result is that viewing produces even more viewing. To put it another way, watching Netflix has a sticky quality. Once you have begun watching, it often takes more effort to leave than to stay.

Perhaps Netflix's greatest impact is that it relies on subscriptions for its revenues, and so it does not carry ads. While there are other streaming services that carry ads and sell products, part of Netflix's appeal is that its shows flow uninterrupted without endless story breaks and a phalanx of ads. Arguably, Netflix has changed the culture so that viewers are now used to watching TV without advertising—and are, as a result, less than enthusiastic about switching back to conventional TV shows chock full of ads. The great irony is that we have a public broadcaster, whose shows are dripping with commercial adverting, and Netflix, which is a business providing its viewers with an ad-free experience.

One consequence of not having advertising is that not only is the story arc different because there is no need to build long pauses and dramatic breaks into shows, but the content is also different. Without pressure to fit stories into 22- or 48-minute time slots, episodes can be longer and, hence, have more sophisticated and complex narratives and greater character development. They are also more expensive to produce, so that shows such as *The Crown, Orange Is the New Black, Stranger Things,* and Marvel's *The Defender* series have an epic and cinematic quality, and they have far more money on the screen than conventional TV shows. The new genre of streaming shows are, in effect, 6- or 8-hour or 12-hour movies. Their structure and screen quality, let alone the scriptwriting, resemble a panoramic movie-watching experience much more than they do a conventional or cable TV program.

The other and far more controversial aspect of Netflix as well as the offerings of other streaming services that don't carry advertising, is that without pressure from advertisers, few of the old rules apply. There are no moral stop signs. Every fantasy is indulged; every horror shown, regardless of how brutal or grotesque it might be; every line can and is crossed. On shows such as *Breaking Bad, Narcos, Ozark, Dogs of Berlin,* and *The Bridge,* everyone, including good people, exist in an abattoir of death and torture. Even by the standards of the dark night of Scandinavian noir, Netflix goes further. The problem for the CBC and conventional broadcasters is that while viewers seem attracted to TV that breaks all the rules, they are—justifiably, in our view—limited in what they can show. As streaming services push the moral boundaries of TV viewing to where they have never been before, public broadcasting remains pleasantly innocent—and perhaps, as a result, more unwatched, at least by a growing segment of the population.

Another change brought by Netflix and the other major streaming services is that they have "globalized" the screen. While cable TV offers many international channels, its subscribers tend to be people who want to watch shows from their countries of origin. Cable can produce viewing ghettos. Netflix original productions have done the opposite. Canadian viewers now routinely watch popular programs from the United Kingdom, Germany, Latin America, France, Israel, Spain, Scandinavia, and elsewhere. One of Netflix's effects has been to internationalize TV to a degree that we have never before seen. This means that Canadian producers have to up their game if they want to compete in this new global marketplace. Happily, Canadian shows such as *Kim's Convenience* and *Schitt's Creek* have already met the test and have gained a global exposure that they might not have had without Netflix.

While Canadian producers have benefited over the years from numerous co-productions with foreign broadcasters, the difference this time is that there has to be more money on the screen, better scripts, more of an epic, movie-like quality. In the face of intense international competition, the old industrial model employed by conventional TV is unlikely to be as competitive as it once was.

Another point is that streaming is reassembling a mass audience. While the history of TV over the last 40 years has been one of increasing fracturing of the audience, some Internet programming has attracted audiences that rival the popularity of programs in TV's golden age. Most critically, such programming has entered the popular imagination and become part of popular culture. These are the shows that people talk about with their friends, recommend to others, and rush home to watch at night. For Canadian TV, Netflix has become a kind of wrecking ball. It not only threatens to erode audiences for conventional and cable TV but has also forced Canadian TV broadcasters to paint on a much bigger canvas if they want to survive.

The problem for Canadian broadcasting policy is that Netflix has become so large and diverse in its offering that it threatens to displace cable. Essentially, for many viewers TV viewing now takes place almost entirely within Netflix and other streaming services. The CRTC's great gamble is that it made cable the main carrier of Canadian content programming. If cable collapses, finding Canadian content will be all the more difficult for audiences.

Streaming has become a crowded place, with multiple services striving to get a piece of the action and cut into Netflix's lead. In 2019, Disney bought out Hulu and initiated two other streaming services, Disney Plus and ESPN+, after it became apparent that its Marvel movies were building Netflix's success rather than its own. Apple TV+ launched in 2019, and YouTube now has two streaming services, YouTube Premium and YouTube TV. Apple alone will launch 30 new series that include a bevy of Hollywood stars in its lineups. There are different streaming services that focus on sports, gaming, Bollywood, television shows from the United Kingdom, horror, and a host of other genres. In Canada, Bell Media's Crave, Videotron's Club illico, Radio-Canada's Ici Tou.TV, and CBC's Gem have joined the mix.

Gem has some advantages. It's free, providing you don't mind advertising (it costs $4.99 a month for a premium service to avoid the ads), and as Johanna Schneller suggests, it's in a position to make bold choices, including developing clear themes and points of view.[35] It's also better late than never, even if the larger ships have long since left port.

While the federal government's Netflix strategy, which we will discuss at some length in chapter 6, seems at first blush like an abandonment of the CBC and conventional broadcasters, the Netflix deal may be one of the only ways for Canadian TV programs to get noticed, both internationally and nationally, in an increasingly crowded space.

The Facebook Dilemma

Here we use Facebook as an example in its own right but also as a stand-in for other social media. What is remarkable about Facebook is that its goals are the exact opposite of public broadcasting. Where public broadcasters try to create public spaces so that citizens can meet and exchange information, Facebook has privatized public space. Information about users helps direct advertising, and users themselves are turned into products.

Facebook threatens the CBC in a number of fundamental ways. First, it has supplanted other media as a principal meeting place where Canadians gather and where a sizable number of them, close to 40 per cent in 2018, get their news.[36] Some 20 million Canadians are on Facebook, and more than 60 per cent check in at least once a day.[37] Second, together with Google, it dominates online advertising to such a degree that all other media are left with little more than scraps. Third, Facebook is now part of the CBC's delivery system. Since Facebook is a principal gateway to the CBC's audience, the CBC can thrive only if its stories, highlights, and promotions are regularly posted and shared. If the CBC's stories don't appear, if they don't make the Facebook cut, then the public broadcaster loses the exposure that it so desperately needs. But appearing on Facebook also has its dangers. The CBC, like other media organizations, risks "brand annihilation" because many users identify posts as just another item that they have seen on Facebook rather than crediting or remembering the original media source.

Fourth, Facebook raises all kinds of issues about the definition and, indeed, the integrity of news. The media giant uses news as a mood modulator rather than as a reflection of the world as it is. In order to keep its users on the site, its algorithms stress feel-good news stories, while downplaying stories that are too grim or depressing. The newsfeed also gives sensational stories greater play and famously customizes news so that users exist in a "filter bubble," which continually reinforces the views that they already have. Facebook also stresses short, visual clips over the written word. Most critically, since 2016 Facebook has been embroiled in a continuing scandal over false news, misinformation, and disinformation, the extent to which Russian trolls and bots systematically targeted certain voting groups with false news stories during the 2016 US presidential election and, in 2019, live-broadcasting (picked up by others and rebroadcast on the Internet) an attack on two New Zealand mosques in which a white-supremacist gunman killed 51 people. In short, Facebook and similar sites challenge the very premise on which public broadcasting is built: in-depth stories, sometimes disturbing truths, accountability news, journalist-driven rather than data-driven stories, and reliable and balanced coverage based on ethical standards.

It is important to remember that while Facebook is often seen as only a single site, its holdings include Instagram, WhatsApp, Messenger, and the virtual reality company Oculus, among other properties. In other words, Facebook's reach extends far beyond Facebook itself. While younger users may be migrating to other services such as Twitch, Twitter, and Snapchat, the Facebook site alone has well over two billion users, and its sister sites also have massive followings. One survey found that, on any given day, users will spend as much time on Facebook as they will in real-life encounters with people. In 2019, the average user spent an astonishing 58 minutes on the site per day, with over two-thirds visiting the site every day.[38] Over half visited several times a day.[39] It also has to be noted that users spend an average of half an hour per visit on Instagram, another Facebook property.

So the image that some observers might have of users buzzing from site to site like bumblebees going from flower to flower couldn't be further from the truth. Facebook is not only the dominant social media site, enjoying far greater popularity than its rivals, with more users getting news from it than from YouTube, Twitter, Instagram, Snapchat, LinkedIn, and Reddit combined, but once in its clutches, many users remain on the site for long periods without drifting elsewhere.

What seems to give Facebook its addictive quality is that there is always something new to check on: new news stories, new reactions, new pictures, new likes, new videos, the latest memes, etc. Sherry Turkle, a professor at the Massachusetts Institute of Technology, believes that we are "being neurochemically" rewarded by an endorphin kick every time we visit Facebook;[40] we are giving our brains the jolt of stimulation that they require. Moreover, to be messaged is flattering—it gives people a sense of belonging and recognition even if they know, at some level, that most of their "friends" are not really friends. But there is also a yo-yo effect when the sugar high wears off. Many users feel let down or that they have wasted their time. Studies have also concluded that watching a continual parade of people that, on Facebook at least, appear to be leading exciting lives, taking great vacations, having wonderful family experiences, leaves many users feeling frustrated and despondent. But what brings users back is that, for many, FOMO—the fear of missing out—is profound and, in fact, intolerable.

In the end, Facebook is about the politics of the personal. As law professor Bernard Harcourt has emphasized in his book *Exposed: Desire and Disobedience in the Digital Age*, people seem to have a compulsive need to put themselves on display, to expose their emotions and purchases, hobbies and passions, to put their lives in a viewing window[41]—in short, "to perform" for their publics. Harcourt compares Facebook to "a mirrored glass pavilion," where users take pleasure in exhibiting themselves, while watching and enjoying the exhibitions put on by others.[42] In the end, Facebook is a "self-presentation" vehicle, a never-ending ad for the person you love the most—yourself. These are "curated identities," where the vast majority of users put their best and happiest face forward all the time, and where even your so-called friends are part of your display, part of your storefront advertising, signalling your popularity and social standing.

Facebook collects extensive data about its users—in effect, getting its most important product for free. It scoops information about its users in at least three ways. First, it famously tracks its users. Once you have joined Facebook, the social network follows you everywhere and never lets go. Second, it tracks non-users who visit sites with Facebook Like and Share buttons, whether people tap those buttons or not. Venture anywhere within Facebook's orbit, and you get sucked into its vortex. Third, according to brand strategy guru Scott Galloway of New York University, Facebook can listen to noise picked up by cellphone microphones. With the aid of listening software, Facebook can "determine whom you are with and what you are doing—and even what the people around you are talking about."[43] Even George Orwell, the author of *1984*, would find this scary.

As mentioned above, Harcourt argues that Facebook's real driver is people's boundless need to perform in front of others. What people don't realize is the extent to which their performances are indelibly inscribed in Facebook's memory. Simply put, the Facebook data machine never stops and never forgets. While the social network famously doesn't sell personal data to others, this information is given to advertisers so that they can target users with highly customized ads. The data has also been given to "partners," including other social media sites, banks, and major corporations.

What's most important about Facebook from the perspective of the CBC is that it has become the world's largest newsroom and a primary source of news for Canadians. One of the key issues raised by Facebook is the worry that its very architecture contributes to online ghettos, or gated communities. First, people whom we friend on Facebook rarely extend beyond a narrow band of family members, friends, co-workers, and friends of friends. The critical point is that our friends are people of our own choosing so that disagreeable characters, distant strangers, and people from other walks of life are unlikely to be invited to the party. Go on virtually any Facebook site, and you'll see a highly segregated world, one in which people of similar lifestyles and backgrounds interact almost exclusively with each other. Harvard law professor Cass Sunstein uses the term "enclave deliberation" to refer to situations where insulated groups speak mostly to themselves.[44]

Compounding the problem is that Facebook's newsfeed is choreographed to reflect the interests of individual users. So instead of the slogan that has adorned the *New York Times* masthead for close to a century, that its job is to provide "all the news that's fit to print," the social network provides users with only the news that its algorithms indicate that they will find appealing. The same filtering process occurs when Google provides information based on your previous searches or when Twitter's "trending topics" are matched to your interests. The danger of this kind of selective exposure, according to Eli Pariser, is that "You get stuck in a static, ever-narrowing version of yourself—an endless you-loop." In the end, "The user has become the content."[45]

danah boyd, a principal researcher at Microsoft Research, believes that this new digital architecture damages public life. As she expressed the problem, "This is super convenient, but it is also seriously narcissistic. What constitutes the public when we're each living in our personalized world? How do we engender public-good outcomes when our tools steer us towards individualism?"[46] Or to put it differently, as Daniel Kreiss asks, how can a democracy function when media is about identity rather than information?[47]

Scholars such as James Webster dispute the notion that the world is divided into identity ghettos because people are still exposed to, and live in, multiple and overlapping media worlds and move from one to another with relative ease. Films, books, comedians, YouTube videos, Netflix films, morning news shows, etc. all expose people to information that they might find uncomfortable. Even on Facebook, opposing views seep through the foliage of opinion in some way.

The argument, however, is not about whether filter bubbles exist but the extent of their influence. Some scholars have concluded that, at the very least, "a sizeable minority of people prefer like-minded news content."[48] Arguably, these are the people who pay a disproportionate amount of attention to politics, but that is debatable. Elizabeth Dubois and Grant Blank argue the opposite—that filter bubbles and echo chambers are less present among people with an interest in politics or those with diverse media consumption habits. Their study of 2,000 Internet users in the United Kingdom found that less than 10 per cent of the population was susceptible to such web-based worlds. They concluded that the heightened attention given to filter bubbles was the result of methodological failure among researchers—carrying out studies that looked only at a single social media site, when Dubois and Blank found that those aged 18 to 34 years of age had accounts on, on average, five social media platforms. Hence, they concluded that this diversity of views undermines the filter bubble–echo chamber argument.[49]

Even if one accepts the argument that the effects of filter bubbles are highly limited, many scholars still believe that some effects exist and there are more than a few scholars who believe that filter bubbles have dangerous repercussions and that they have helped societal divisions and hyper-partisanship reach dangerous levels. Cass Sunstein argues that the problem is that "choices that seem perfectly reasonable in isolation may, when taken together, end up disserving democratic goals" and that there are dire risks associated with "any situation in which hundreds of thousands, millions, or even hundreds of millions of people are mainly listening to louder echoes of their own voices."[50]

Yascha Mounk believes that democracies are "de-consolidating," in part because extremist views now have a place to incubate and spread.[51] By empowering outsiders and authoritarians, by giving them legitimacy and a launching pad, social media has helped destabilize democracies. Old and venerable institutions are bypassed, mocked, and under attack, often using fake news.

To some degree, the argument for public broadcasting rests on its ability to create common meeting places, to be the public square, where the issues of the day can be presented and deliberated. As we have seen, the CBC once played that agenda-setting role. The question is whether, in this new climate of hyper-division and fragmentation, the public broadcaster can still play this role. If it can't, then what role can it play?

Another challenge presented by Facebook is that the social network has redefined the nature of news and, indeed, all media. James Webster believes that Facebook has transformed media into a highly choreographed system of "mood management."[52] The goal is not to give users the news that they will need to understand the forces that are shaping their lives but to satisfy desires and provide the sugary treats that will make users return to the site. Facebook is "not only optimized for engagement" in a general sense but also knows which signals individual users will respond to. For instance, the social network has learned through continual monitoring, focus group testing, and experiments conducted on large samples of users that happy and positive stories are more likely to go viral. Its news managers, who use algorithms designed by engineers, also know whether individual consumers prefer text, video, or images and in what combinations; the keywords and shades of colours that they respond to; and whether they are likely to want to read about some topics and not others. Facebook's News Feed is also based on how often users interact with certain friends, the websites that they have visited, and the number and length of previous visits. As Kate Losse observed, "Because Facebook is a business, what they believe you should see is based on what is good for Facebook."[53]

The problem is that Facebook's journalistic formula has spread across the Internet. Franklin Foer, a former editor of the *New Republic*, has described the challenges that he faced having to compete against news organizations whose editorial policies were based on data analytics. As Foer recalls, "The subject could be serious, the presentation had to be fast and fun, geared to spread via Facebook."[54] Negative stories didn't sell, and because of tools such as Chartbeat, he could see "the flickering readership of each and every article" in real time. To survive, "We simply had to mimic the rest of the internet—write about the same outrage as everyone else, jump on the same topic of the moment. Clicks would rain down upon us if only we could get over ourselves and post the same short clips from the Daily Show as everyone else." As we will discuss in the next chapter, this is a more or less accurate description of how CBC News now operates in producing online stories.

The danger to Foer "isn't just the media's dependence on Silicon Valley companies. It's the dependence on Silicon Valley values. Just like the tech companies, journalism has come to fetishize data.... Reporters and journalists can assert otherwise. They can pretend to rise above the information, to selectively ignore the numbers and continue the relentless pursuit of higher truths and nobler interests. But data is a Pandora's box. Once journalists come to know what works, which stories yield traffic, they will pursue what works. This is the definition of pandering and it has horrific consequences."[55]

Another danger is that neither Facebook nor Twitter, for that matter, requires proof of identity. This has left them open to tens of thousands of imposters, parody accounts, and automated bots that produce and amplify an endless barrage of misinformation. It's a world in which ghost bots using real names and photos are armed for information warfare. As mentioned above, Russian trolls and bots unleashed an industrial-scale disinformation campaign during the 2016 US presidential election that contributed to suppressing voter turnout among African-Americans, Hispanics, and women.

The challenge for public broadcasting is that the Facebook news model is the very opposite of what public broadcasting should be. While news can sometimes be light and breezy and make people feel good, it also has to be tough-minded, tell painful stories, and tackle issues that need more time to explain because they are complex and many-sided. The ultimate Facebook problem may be that it fools people into thinking that they are getting the news, that they have a handle on the world, when the stories they are reading are only those that they like served up in an easy and comfortable way.

As mentioned above, further exacerbating the CBC's Facebook problem is that when its stories or promotions do make it onto Facebook's News Feed, they run the risk of brand annihilation shared with virtually every other media organization. While having their stories or segments go viral is usually beneficial for media organizations, the fear is that those items can be lost in the endless shuffle of News Feed stories and become just another story that people credit to Facebook. The problem is that the CBC can't escape the iron grip created by Facebook and other social media.

Can the CBC Survive the Deluge?

One of the CBC's basic dilemmas is that there is now too much media and too many claims on the audience's attention. While the CBC faces blistering competition in every area of music, sports, drama, entertainment, and news—competing against media companies that offer a glittering kaleidoscope of programming options—the CBC concentrates its attention in two key areas: Canadian and local programming. By competing everywhere, it loses the capacity to refocus and build its strength in areas where it can make the greatest difference. This is especially the case given the impending collapse of the newspaper industry in Canada and the increasing turbulence facing local TV stations. Basic news reporting at legislatures and city halls, at school boards, and about the economy is likely to be in short supply in the years ahead. The CBC will have to reoccupy arenas that it has largely abandoned—remember its cutbacks to local TV stations and to its radio programming? Winning back audiences may not be automatic. In addition, while tech media tends not to be interested in local news, companies such as Google, Reddit, and even Facebook are beginning to smell opportunities for local expansion. Moreover, the CBC will need the FAANGs to access its audiences. The relationship that it carves out with these new portals and hosts will determine much of its future. None of this will be easy.

The great irony is that even though the CBC may not be able to compete against companies such as Disney and Comcast, Facebook and Netflix, it provides a solution to some of the problems that they have created. Where tech media giants such as Facebook exacerbate social and political divisions by creating ideological bubbles and ghettos, the goal of public broadcasting is to create large public squares where everyone can meet. Where the FAANGs are largely in the data-collection business—collecting, curating, auctioning, and selling information about their users—public broadcasters, aside from having rudimentary data on their audiences, do not keep such information and are therefore unlikely to turn the private lives of their audiences into a product.

Then there is the issue of fake news and disinformation. While Facebook had few defences against the wholesale importation of disinformation and misinformation during the 2016 US presidential election, public broadcasters produce their own news and use old-fashioned gate-keeping standards overseen by people, not algorithms. In an age when it is difficult for citizens to know what news to trust, the CBC can be a reliable old friend, but its pursuit of clicks is corrosive because it undermines its ultimate value in searching for and highlighting facts in a media universe drowning in opinion. Moreover, where users are burdened by the increasing weight of subscriptions, micropayments, and pass-through charges, and where, therefore, for many, their media worlds—what they can afford—are becoming smaller rather than larger, the CBC's main channels are freely accessible. High-quality information may be the last bastion of public broadcasting, its strongest selling point.

Chapter 5

The Collapse of Sports and News

This chapter examines the CBC's loss of *Hockey Night in Canada (HNIC)*, its retreat from big-time sports, and its increasingly precarious hold on its news audiences. In the previous chapter, on the impacts being made on audiences by social media sites such as Facebook and streaming services such as Netflix, we made the case that CBC drama and entertainment programs are losing the battle for eyeballs in the new attention economy. These very same forces are upending relationships in every genre and across every platform. Here we examine the CBC's positioning in two crucial areas that have been integral to its stature and credibility. Sports has long been the heartland for the CBC as it has attracted the largest audiences and helped place the public broadcaster at centre stage of the Canadian cultural experience. News is the very lifeblood of public broadcasting. If the CBC becomes just another news voice in an increasing cacophony of news choices, if its news presence slowly evaporates, there is little future for public broadcasting. The CBC's battle for survival will have been lost.

The Painful Loss of *Hockey Night in Canada*

Losing *HNIC* may have been the best thing that happened to CBC in a decade. That's not just because it solved a big, short-term problem for the broadcaster, even though it certainly didn't look that way at the time. Rogers Communications stunned the sports and media worlds in Canada in the summer of 2013 by announcing a $5.2 billion, 12-year deal with the NHL. By outbidding both the public broadcaster and its specialty-channel rival TSN, owned by BCE, Rogers gained control of the NHL broadcast rights on conventional television through its Citytv network and its Sportsnet specialty channels, on Rogers radio stations, on the Internet, and on mobile devices.

Perhaps the real surprise in the CBC's failure to negotiate the renewal of its long-standing contract with the NHL was that it was surprised that it lost the bidding war. CBC executives apparently thought that a deal with the NHL would happen almost automatically because the CBC had always had the rights to broadcast NHL games. They appeared not to have anticipated that the world had changed dramatically and that renewing the NHL deal was, at best, only ever a long shot. Changes in Canadian media over the preceding three years meant that the CBC simply couldn't outbid private broadcasters for high-priced programming rights if the privates thought that they could make money on the programming.

Author David Shoalts argues that the CBC missed two golden opportunities to retain *HNIC* in some form.[1] First, the NHL was irritated by some of the commentary on CBC's version of the program, particularly by what it saw as the antagonistic relationship between host Ron MacLean and NHL president Gary Bettman—Bettman announced at one point that he would no longer appear on the show after being roughed up by MacLean in an interview; nevertheless, the NHL offered the CBC two games featuring Canadian teams on Saturday nights. The CBC would have to pay roughly double what it had paid in the previous contract, and the NHL could broadcast other games at the same time. The CBC didn't like the deal but didn't have time to reconsider. The NHL soon realized that what it really wanted was a "gatekeeper": a single broadcaster that could act as a broker, distributing rights in Canada to the highest bidders. This is precisely what Rogers intended to do.

A second opportunity for the CBC arose in its negotiations with Rogers. Although Rogers did not have the stations needed to cover the country in the same way that the CBC did, the CBC decided to capitulate instead of bargaining for a better deal. As Shoalts describes the CBC's failure, "Nobody at CBC had the foresight to say, 'Okay you guys tell the entire country *Hockey Night* is off the air and by the way we know you need us because your channels can't be seen across the country. Nobody had the guts to do that.... The CBC is horribly, irreparably damaged and nobody had the balls to say [to Rogers] if you won't give us a better deal then you're taking the heat."[2] As we shall soon see, Rogers effectively sacked the CBC, with barely a whimper in response.[3]

It was at the time widely perceived as a disaster for the CBC. Hockey had long been a symbol of national identity and belonging. Legendary CBC personality Peter Gzowski once remarked that "hockey was the common Canadian coin,"[4] and literary scholar Jason Blake has pointed out that while "most Canadians do not wake up thinking about hockey

... the sport remains a comfortable confirmation of the nation's existence."[5] As the broadcasting home to the "home game," first on radio, then on television, and later online, the CBC had been at the epicentre of Canadian sports culture since 1936. The public broadcaster would lose the goodwill that went with the audience of almost two million that regularly watched the first of the weekly Saturday HNIC doubleheaders[6] and the more than half a million that tuned in to the second game, usually broadcast from western Canada. Hockey was the only program that many of those, including younger audiences, watched on the CBC. Now they would have no reason to pay any attention to the public broadcaster. The accepted wisdom was also that, without Saturday night and playoff hockey, the CBC would lose millions in advertising revenue. In fact, the lobby group Friends of Canadian Broadcasting estimated that HNIC accounted for half the CBC's ad revenue and 30 per cent of its viewers.[7]

Starting in the fall of 2014, CBC would become a bit actor in a show run entirely by Rogers. HNIC doubleheaders would stay on CBC on Saturday nights until 2018, and CBC would carry playoff games as well. The initial four-year deal gave the CBC time to figure out how to replace hockey. It decided to do nothing. In late 2017, its arrangement was extended for the remainder of Rogers's 12-year deal with the NHL.

For the full 12 years, Rogers would choose the on-air talent and program content, produce game coverage, and pocket all advertising revenue. CBC had to provide Rogers with studio and office space and pay the salaries of the staff who produced the show. To add insult to injury, the CBC did not receive funding to pay for the severance packages of those employees who had to be let go because of the deal. It now had to go to Ottawa to ask for the severance money.

The public broadcaster had been reduced to a supplicant, a distributor of a private broadcaster's programming with the rights to "stream all nationally broadcast NHL games on the CBC Sports app, the forthcoming 'over-the-top' CBC TV app [Gem] and on the CBCSports.ca website."[8] As Scott Stinson noted in the National Post, "All the talk about the public broadcaster's higher purpose is undercut rather a lot when it is gifting its airtime to a competing private conglomerate, no matter how many promos for the Baroness Von Sketch Show that Bob Cole has to read on Hockey Night."[9] HNIC had become simply, to borrow Ken Goldstein's description, "Rogers on CBC."[10]

But it was a much greater surrender than just turning CBC airtime over to a private broadcaster as Rogers is much more than a broadcaster. As mentioned previously, together with Bell Canada, it owns MLSE, the sports team and real estate conglomerate that in turn owns the Toronto Maple Leafs, the NHL's most successful franchise and hockey broadcasting in Canada's main audience draw. MLSE, which also owns the Toronto Raptors of the NBA, the Toronto FC soccer team, the CFL's Toronto Argonauts and the American Hockey League's Toronto Marlies, is the kingdom and power of Canadian sports. On its own, Bell owns 18 per cent of the Montreal Canadiens, and Rogers owns the Toronto Blue Jays. Given its ownership position, Rogers has a vested interest in promoting both the Leafs and the NHL.

Sports team ownership is part of a larger corporate strategy for Rogers and BCE. Sports programming was seen as the key to revenue growth for Rogers and BCE because it was a way of growing their wireless and Internet subscriber numbers. The goal was to use their ownership of the exclusive rights to televise hockey, baseball, football, basketball, and other sports to encourage customers to subscribe to Internet or wireless services if they wanted to watch sports on anything more portable than a TV screen.[11] Sports was also seen as a way to keep viewers from cord-cutting.

By turning over all editorial control of *HNIC* to Rogers, the CBC was now a participant in promoting not only the NHL, which it had always done, but also Rogers and its corporate interests. Editorially, the CBC is now a part of what Christopher Waddell has described as a "hall of mirrors." For example, interviews with a Maple Leafs player would be done by a television host employed by Rogers, the interview would take place in a building partly owned by Rogers (now Toronto's Scotiabank Arena but formerly Air Canada Centre), and broadcast on a Rogers channel on a program directed and produced by Rogers employees. As he noted in *How Canadians Communicate V: Sports*, "Reportage by media affiliated with teams themselves is inevitably compromised, caught in conflicts of interest in which the mirrors shape and distort coverage, turning journalism into promotion and public relations."[12]

In such a world, there isn't room for detailed reporting on concussions or use of painkillers by hockey players or questioning any aspects of league decisions. In the event of any future labour disputes with the NHL Players' Association (of which there have been four in the past 16 years), it is clear which side Rogers and the CBC would be on. It was the equivalent of handing over editorial oversight and control of all aspects of CBC election campaign coverage to the party in power and a sad commentary on how sports journalism has devolved into sports promotion for commercial enterprises, hardly the mandate envisaged for public broadcasting either in the past or in the attention economy.

Despite the existential journalistic questions that the Rogers *HNIC* deal raised for the CBC, continuing to carry hockey actually solved a critical, short-term problem for the public broadcaster. Had *HNIC* been yanked from CBC's schedule starting in October 2014, finding replacement, Canadian-produced programming to cover the six hours of air time eaten up by two games every Saturday night would have been difficult and certainly hugely expensive either to buy or to produce. *HNIC* counted as Canadian content, allowing the CBC to continue to claim that it was still the great champion of Canadian content on TV. The added cost of producing new programming was the last thing the CBC needed in 2014, when it was in the midst of implementing a $115 million budget cut imposed by the Conservative government. So not only did the CBC lack the money needed to compete against Rogers for the rights to broadcast NHL games, it also lacked the financial resources to air Canadian programs that might be broadcast instead of hockey. Another dilemma was that even if the CBC had been able to mount Canadian programming to replace *HNIC*, those shows would have had to go head to head against *HNIC*. Audiences would have been minuscule.

Broadcasting the Rogers-owned *HNIC* also allowed the CBC to maintain the illusion that it was still in the sports game. As mentioned in the introductory chapter, the loss of *HNIC* is part of the CBC's wider retreat during the preceding decade from sports broadcasting. The CFL, curling, and soccer, including the FIFA World Cup, left the CBC for specialty sports channels long ago, and basketball never seemed to make it onto the CBC's radar—a critical misreading of the sports audience, particularly younger viewers.

What's left for CBC Sports is the Olympics every two years, around which the broadcaster has added the rights for the Calgary Stampede rodeo, Canadian Premier League soccer, the IAAF Diamond League Track and Field Championships, and FIVB Volleyball in what is described as a pivot to "high-performance" sports.[13] Much of that is broadcast online, where it draws small audiences, and at best presenting only highlights on television. In short, the CBC is stuck with the leftovers that the private networks don't want because audiences and advertising revenue are simply too small. Big-time sports have moved elsewhere.

The Great Canadian Sports Gamble

As part of any discussion about sports broadcasting, some background on the recent history of the Canadian media is necessary. In truth, the writing was on the wall for CBC's involvement with professional sports from the time of the bankruptcy of Canwest Global in late 2009. Buried under a mountain of debt that it could not repay when faced with advertising starting to move away from mainstream television, Global came apart at the seams. Its collapse marked the visible and painful failure of convergence—a consolidation strategy pursued aggressively by Canadian media owners over the preceding decade and a half.

In theory, convergence would maximize the profits from advertising for media owners when, in the late 1980s, the Conservative federal government lifted restrictions that had prevented media cross-ownership in individual communities. Now the same parent company could own, often in the same market, conventional television stations, newspapers, radio stations, and the related websites for all of them as well as specialty television channels. This meant that, in some markets, the CBC was the only other voice, the only alternative to media owned by just one or two media conglomerates. Moreover, as the CBC was limited to broadcasting, the policy of allowing cross-media consolidation had the perhaps unintended consequence of reducing the CBC's reach and audience.

Under convergence, profit would supposedly be maximized by taking content produced for one medium and displaying it across all the others and creating "one-stop shopping" advertising packages that allowed advertisers access to multiple platforms. Initially, that was primarily true for news and information. The reporters and editors working for a newspaper, for instance, could repurpose their print stories for distribution on their sister television or radio stations or their websites.

Media owners took advantage of this new freedom to consolidate. Québecor had moved aggressively in this direction starting in the late 1990s, buying the English-language Sun newspaper chain to add to its ownership of *Le Journal de Montréal* and private, French-language television station TVA. It later added cable and wireless company service provider Vidéotron and the Osprey newspaper chain outside Quebec. Bell Globemedia (later reorganized as CTVglobemedia) brought together the *Globe and Mail* newspaper and CTV, Canada's largest, privately owned television network under the same roof as well as the specialty channel and wireless holdings of BCE. Canwest Global added the former Southam newspaper chain, purchased from Conrad Black, to its national Global television network and specialty channels.

Convergence was a flawed strategy on almost every level. Advertising soon began the slow transition away from mainstream media, starting with Craigslist, Kijiji, and others taking over the lucrative classified-advertising market from newspapers. With advertising revenue starting to decline by the mid-2000s, it became tougher and tougher for media conglomerates to maintain their traditional 20 per cent or higher rates of return that investors and owners had come to expect (rates of return that are hard to imagine today) for newspapers and television stations. Yet the conglomerates still had to make the required interest payments on the money they had borrowed to put the converged entities together. In short, they bit off far more than they could chew.

But once committed to convergence, the train could not be stopped. The theory again was that each entity—whether it be a newspaper, website, or radio or television station—would need fewer editorial workers if it were sharing the same content with all its other related entities. News organizations could safely lay off reporters and editors and still produce material for all their media platforms, and that's what happened. By 2005, the elimination of editors, producers, and journalists had started in earnest, and more than a decade later, the layoffs, voluntary buyouts, and newspaper and television closures were still in full swing, leading some to ask how long it would be before there was no one left to cut. By 2018, almost every newsroom in the country was running on skeleton crews, and as mentioned in the introduction, some, particularly small enterprises, had shut down, while others were on the brink of collapse.

Convergence may have made sense for accountants on spreadsheets, but the reality was a disaster on every level. Newspaper reporters couldn't be turned into television reporters—the skills are different. While some can be multi-skilled and therefore multi-tasked, very few could juggle multiple media requirements on the same day. Most couldn't make that transition easily or didn't want to do it. They liked what they were doing in their own medium and saw no reason to change.

Just because a newspaper and its TV network were owned by the same parent company didn't mean that they would get along. Why, for example, would the *Globe and Mail* give one of its exclusive stories to CTV to run on its national newscast the night before the newspaper hit the streets? Journalists at each jealously guarded their stories, and sharing was the rare exception. That was equally true for Canwest's newspaper and television divisions.

All this came to a head with the sharp contraction of advertising by consumer retailers, automobile companies, real estate firms, and the financial services sector in and after the global financial meltdown of 2008–09. Although Canada sidestepped the worst fallout from a recession felt in many countries, it didn't escape it entirely. Canwest Global didn't escape it at all.

Its collapse consigned convergence to the scrap heap, a complete failure, but not before costing billions of dollars and thousands of jobs. Bankruptcy trustees split up Canwest Global, and a new company, Postmedia, bought the newspapers. Western Canada cable and specialty channel provider Shaw Communications purchased the Global television network, with its over-the-air and specialty channels, offloading it in 2017 to Corus Entertainment, its related media company. The *Globe and Mail*–CTV relationship also fell apart. BCE took CTV, and the Thomson family bought back 85 per cent of the *Globe*, subsequently adding BCE's 15 per cent interest. Not to be left out, Rogers Communications took over the CITY-CHUM network of television stations, integrating them into its extensive radio network, cable and wireless operations, and specialty channels.

The convergence nightmare was not the only factor that was leading to the disarray and decline of Canadian media. Internet advertising had begun to claim its print and broadcast victims, and Google and Facebook would eventually eat up 80 per cent of online advertising in Canada, leaving conventional newspapers and TV stations with the leftovers. In addition, the attention economy would begin to hit with torrential force as the number of competitors in every area multiplied by orders of magnitude.

Convergence was dead except for the grievous debt that it left behind, but like a phoenix, vertical integration and the new subscription economy rose to take its place. Instead of cross-ownership by one parent of different media within the same market, under the new strategy one parent company would now own the producers of content and its distributors across cable and satellite systems, on the Internet, and on mobile devices. Vertical integration differed from convergence in that it would be consumers, not advertisers, who would be the main sources of revenue. Vertical integration depended on the new subscription economy, in which consumers paid for cable, satellite, and/or Internet television packages; related monthly fees for specialty channels; plus monthly home Internet service, mobile phone and tablet plans, and data charges.

One can argue that vertical integration was part of the CRTC's unstated national champions policy, which encouraged Canadian communications conglomerates to develop the resources needed to compete against foreign competition, invest in new infrastructure, and produce Canadian-content programming. It is worth noting that, as mentioned earlier, while the private media giants were encouraged to grow larger, the CBC was allowed to stagnate, to the point where it was unable to compete against the private media behemoths when they were after the same prize. The CBC simply did not have the money to bid against private broadcasters for anything the privates really wanted to buy. Both Rogers and BCE, through Sportsnet and TSN, desperately wanted the NHL.

Against either of these major players, the CBC didn't have a hope. Rogers or TSN could recover the costs of purchasing the rights to NHL hockey and more by increasing the number of subscribers to its specialty sports channel packages (which, with a larger subscriber base, would concurrently increase what they could charge on their channels for advertising); by persuading mobile phone subscribers to switch to its service and perhaps switch to its Internet and/or cable, satellite, or Internet TV service as well; and by selling NHL games watched on computers or wireless devices. It could also build specialty online-content packages for which it could charge NHL fans even more.

The key to this was the then confident knowledge among broadcasters that there existed an almost insatiable appetite for live sporting events—or so they thought. Every one of the top-20 largest TV audiences in Canadian broadcasting history was for a sports event.[14] Out of the top-100 most highly rated shows on US TV in 2017, 62 of them were sports telecasts.[15] Sports broadcasting brought in close to 40 per cent of all TV advertising revenue in the United States in 2015, a simply staggering number.[16] Sports also drives social media traffic. According to audience monitoring firm Nielsen, more than half of all discussions about TV programs on Twitter are about sports.[17] Simply put, sports was a gold mine, a broadcasting bet that had paid off handsomely time and time again. Moreover, live sports looked like the perfect strategy to prevent cord-cutting. No one wants to watch Saturday night's NHL game the following Tuesday evening. Once fans knew a game's final score and had watched the highlights on a sportscast or their mobile devices, what was really left to see? Broadcasters also thought that, unlike other types of viewing, where people would multi-task, flit from channel to channel or binge-watch programs on a streaming service, sports fans were more likely to watch entire games and were more likely to see the ads.

Hockey, the broadcasters believed, would also prove a huge attraction in the new pick-and-pay satellite and cable television world, where consumers could buy individual channels rather than having to purchase a bundle of channels, many of which they never watched. The sports channels could be sold at a premium, with the owners getting more revenue than they did in the bundled-channel era. They could also market the sports channels individually to mobile subscribers as a separate monthly service, adding data charges to their customers' bills if they watched a lot. Hockey would drive all this, and advertising revenue on all these platforms would be the gravy on top of the core subscriber fees. In recounting, it sounds a lot like the same theory that drove convergence.

By contrast, the CBC could cover the costs of buying NHL rights only through selling advertising. It couldn't raise revenue from cable or satellite services, specialty channels, mobile phone or tablet subscriptions, or data charges or by providing monthly Internet service to homes because it had no presence in those worlds. Compared to Bell and Rogers, the CBC had more than both hands tied behind its back when bidding to renew its NHL rights or for any other sports broadcasting rights. It didn't stand a chance of matching what the vertically integrated could offer.

Before the House of Commons Heritage Committee in October 2016, CBC Executive Vice-President of English Services Heather Conway tried to make a virtue out of the public broadcaster's plight, stating, "To suggest we lost hockey, like we screwed up somehow, is actually not a fair characterization. The cost of professional sports rights is through the roof. It's not a great use of taxpayers' money, and that's by and large why we are out of the big, expensive professional sports rights."[18] If that's what CBC management really believed, why did the CBC try to hold on to the NHL, and why does it now go through the charade of broadcasting *HNIC*? That sounds more of an *ex post facto* rationalization of failure.

The question at the end of the rainbow is whether the BCE and Rogers great sports gamble will pay off. There is mounting evidence that the glory days of sports broadcasting are quickly fading and that sports broadcasting is now in what the research firm MoffettNathanson has called a "structural decline."[19] While NFL football is still a prime mover of the TV schedule in the United States, accounting for 46 of the top-50 most-watched programs in 2018, the league lost 17 per cent of its audience between 2015 and 2017.[20] Observers speculate that the politics of the national anthem, concussions and player safety, and moving teams in and out of key markets were all factors in the decline. Viewership rebounded slightly, by 5 per cent from the previous low, in 2018. The 2018 Super Bowl, however, registered an audience of 103 million viewers, a nine-year low.[21] In the first part of 2018, attendance at Major League Baseball games plummeted to its lowest number in 15 years, with average attendance below the 30,000 mark.[22] TV audiences for the World Series seem to be on a steady decline, prompting widespread concern that "baseball is dying." In fact, audiences for the 2018 World Series between the Boston Red Sox and the Los

Angeles Dodgers—in two prime baseball markets—was the fourth-least-watched series since the beginning of TV—down 25 per cent from the previous year.[23]

Basketball also has its troubles. While the NBA has a younger fan base than other sports, its games are sellouts, and it has a global audience, viewership declined by a grim 22 per cent on TNT and 5 per cent on ESPN in 2018.[24] Average viewership in the United States for the NBA finals has declined since 2015 and has stagnated to the point where it is roughly the same as it was in 2004.[25]

Unfortunately for Rogers, hockey has been a mixed blessing. Much to its horror, the size of Canadian TV audiences during the playoffs can swing wildly from year to year, depending on whether Canadian teams are playing. The failure of any Canadian teams to make it into the 2016 NHL playoffs meant that first-round playoff-game viewing that spring was a stunning 55 per cent lower than the year before. Second-round games were, on average, down 52 per cent, while third-round games fell 11 per cent from 2014 to 2015.[26] *HNIC* audience numbers bounced back somewhat in the 2016–17 season as Rogers abandoned many of the changes that it had made to the program in favour of returning to what was essentially the CBC style of game coverage, complete with former CBC host Ron MacLean and his small-town folksiness replacing the more urban and hip George Stroumboulopoulos.

That strategy may be paying off as, in 2018, the Winnipeg Jets versus Las Vegas Knights semi-final series attracted increased audiences from the year before, as did the final round between the Knights and the Washington Capitals. A year later, in the spring of 2019, it was back to the misery of 2016 for Rogers. Despite Calgary, Winnipeg, and Toronto being considered Stanley Cup contenders, they were all eliminated in the first round of the playoffs, leaving no Canadian teams in post-season play. This had the obvious negative impact on audience levels on television and on mobile devices: viewership on some nights was as low as under 400,000.

But it was not all bad news for Rogers as its Sportsnet channel also carries Toronto Raptors NBA games. The Raptors' late-May 2019, game-six, Saturday night victory over the Milwaukee Bucks, which sent the team to the NBA finals for the first time, set a basketball television audience record in Canada. It was watched by 3.1 million viewers, on average, with a peak of 5.3 million.

By comparison, playoff hockey with no Canadian teams struggled to find viewers. "The most recent Numeris Top 30 Live+7 totals featured three Raptor games split between Sportsnet and TSN," noted Bill Brioux in late May 2019. "On the hockey side, in mid-May, not one NHL

Stanley Cup game on either CBC or Sportsnet cracked the May 6–12 list of top-rated shows in English Canada."[27]

Even when there are increased audiences, that doesn't necessarily translate into increases in advertising dollars. According to Ken Goldstein's calculations, advertising on *HNIC* has dropped precipitously from when the CBC controlled the show.[28] In 2012, when the CBC was in charge, the advertising haul was $122.7 million. In 2017, Rogers on CBC brought in just $95.3 million.[29] One has to keep in mind that *HNIC* is just one part of a much wider Rogers hockey package that includes broadcasts on Rogers Sportsnet, the selling of French-language broadcast rights, and serving mobile and Internet customers.

Hockey's problem, however, is that the average age of hockey fans rose to 49 in 2016 from 33 in 2000.[30] Millennials are quickly disappearing from the stands and from screens. Interestingly, US TV ratings for the Stanley Cup playoffs, while varying greatly from year to year, have barely moved the dial since 2008. NBC attracted an average viewership of 4.6 million in 2008 and 4.8 million in 2018.[31]

One exception to this general decline is viewership for the Olympic Winter Games—at least in Canada. For Canadian audiences, the Olympics have taken on a special meaning. Largely because of the country's exceptional recent performances in the Winter Olympics, the games have become a kind of national spectacle in which Canadian athletes and the country itself basks in celebration and self-congratulation. According to ratings tracker Numeris, during a five-day period in February, almost two-thirds of all Canadians tuned in to CBC/Radio-Canada's coverage of PyeongChang 2018, with a total of 23.4 million watching at least part of broadcasts across all English- and French-language television network partners and digital streaming simulcasts on the CBC's Olympics websites and its apps.[32] This is a massive audience by Canadian standards and one of the few contemporary instances in which Canadians have gathered together to watch the same events.

CBC has decided to remain with the Olympics. After losing the bidding for the 2010 Vancouver Winter Olympics and London 2012 summer games to CTV, CBC won the rights for the winter games in Sochi, Russia, in 2014; Rio de Janeiro in 2016; and the next four games, both summer and winter, until 2024. Barry Kiefl, president of Canadian Media Research Inc. and the director of research for the CBC from 1983 to 2001, complains that, unlike other broadcasters, the CBC refuses to say how much it paid for Canadian Olympic rights and that, in financial terms, the Olympics are a losing proposition. But unlike in hockey, the CBC did not face serious bidding competition from Rogers or BCE,

both of whom seemed happy to take some Olympic events on their specialty sports channels Sportsnet and TSN, offloaded from CBC, without
the hosting expense and schedule disruption to their broadcasts of professional hockey, football, baseball, and soccer that would come with
wall-to-wall coverage of the games.

The Olympics have not done as well with American audiences.
NBC's broadcast audience for the 2016 Summer Olympics in Rio was 18
per cent lower than the 2012 games in London, even though Rio's time
zone was much more favourable to American viewers. It was the first
games since 2000 in which the NBC audience had declined.[33] Two years
later, it was even worse for NBC as the Winter Games in PyeongChang
were the least-watched Olympics in NBC history.

While sports seem to have lost some of their glitter, sports specialty
channels remain a profit centre, even though it isn't a complete surprise
that sports-channel subscription numbers are falling as well. Between
2011 and 2015, Rogers Sportsnet lost about a million subscribers, while
BCE's TSN lost about 200,000,[34] although just under ten million homes
across Canada still subscribe. In the United States, ESPN, the specialty
channel sports leader owned by Disney, had about 86 million subscribers in mid-2018, down from 100 million at the end of 2011.[35]

The decline in sports viewing on television may have a lot to do with
the nature of the new attention economy. To put it differently, the same
forces that have led to a dramatic decline in almost all aspects of CBC
viewing are also having an effect on sports broadcasting in general. First,
there can be little doubt that the doping era in baseball and other sports
and long-term brain damage from concussions in football and hockey
have taken a toll on the reputation and legitimacy of sports generally. In
the case of hockey and football, they have led to fewer parents enrolling
their children in these sports. In the end, fewer players means fewer fans.

In hockey, in particular, the refusal of old-time hockey executives to
be harsher on hits to the head and the practice of finishing checks, as
well as continuing to tolerate some fighting, even though it is declining, have left more than a few observers and fans bewildered. Other
factors, such as the expense of attending games, the time crunch experienced by most families, and the priority given to elite performers at the
expense of ordinary children who just enjoy the game, have all taken a
toll. As a by-product of that decline among younger viewers, e-sports
is experiencing an extraordinary boom. In fact, viewing of e-sports—
people watching others play computer sports games—skyrocketed by
more than 25 per cent in 2018 alone; one estimate has more than 20 per
cent of male millennials now watching e-sports.[36]

Observers argue, however, that the main reason for the decline in sports viewing is that the Internet has both changed the nature of viewing and created massive saturation at the same time. In other words, sports are losing the battle for dominance in the attention economy. Susan Jacoby has argued that watching sports on TV is now an interrupted experience, an experience in partial attention.[37] Fans have lost the capacity to sit through entire games. Some 80 per cent of fans around the world, according to Nielsen, are doing other things online at the same time that they are watching a game on TV or in the stands.[38] As is the case with so much other media, meals have given way to snacks, so that younger fans increasingly consume sports in short bursts through updates, highlight packages, fantasy league prompts, Twitter storms, and Facebook posts rather than by watching full games. Jacoby has noticed, in observing baseball fans, for instance, that, in the absence of spectacular home runs or high-scoring games, fans think that nothing is happening.[39] They lose patience when a game slows down and don't see or appreciate the more subtle, tactical battles that are being played out in front of their eyes. To Jacoby, these tactical moves are the heart of the game.

At the same time, sports broadcasting has reached beyond the saturation point. Sports leagues have upped the number of televised games so that there is always a game to watch, look forward to, or reminisce about, and even player drafts have become major spectacles, with speculation and hype building for weeks before the drafts themselves. There are now so many sports channels, league and team websites, social media platforms, blogs, podcasts, YouTube channels, fantasy leagues, and streaming services that sports are everywhere, inescapable, and always on. In part, this phenomenon has been fuelled by the expansion of sports gambling, to the point where it is now socially acceptable and legal in many places. TSN alone has no less than five specialty sports TV channels, and Sportsnet has four. There is yet another sports channel, run by the French-language network TVA. *Bleacher Report* now has an Instagram account that includes "House of Highlights," an ad-free, continuous-highlight reel. YouTube's sports-highlight channels have massive audiences, and *ScoreStream* and *Overtime*, which are aimed at local college and high school sports, churn out almost non-stop highlights. In terms of sports streaming, DAZN has moved aggressively into the Canadian space, signing deals with Shaw, Rogers, and Bell and buying the rights to many smaller sports events, which tend to generate niche audiences. But DAZN still ranks far behind Amazon's Twitch and ESPN+ as a major force in the streaming universe.

One can argue that what happens to sports broadcasting will dictate much of what happens to Canadian media generally. Looking back, the CBC was lucky to have lost *HNIC* to Rogers and had little choice but to retreat from big-time sports. Sports broadcasting is now saturated, rights are too expensive to bid for, and audiences and advertising may be slowly disappearing, and it was only through advertising that the CBC could have covered its costs for the NHL rights. The cost of keeping *HNIC* and contending for the rights to other sports would have meant cutting costs across other CBC services and programming. This might have endangered the entire network. Yet the blow from the loss of *HNIC* to the CBC's stature and prestige was real, and so also is the loss of a bridge to younger audiences that came with it.

It's also not clear that the great gamble on sports ownership and broadcasting that Canada's private broadcasters have made will succeed. What is clear, however, is that the success of Canadian media generally may depend on what has become an increasingly risky roll of the dice.

Adrift in a Digital Sea: The Precarious Future of CBC News

The same forces that have limited the CBC's ability to compete in sports have affected its ability to compete in news. Decades of budget cuts have combined with the challenges brought by digital change to transform CBC News from a news powerhouse, with the ability to set the national media and public agendas, to just another voice amid a vast chorus of other voices. News programs and bureaus are understaffed, specialized reporters are vanishing, and investigative work seems to have less of a priority.

At the same time, many, if not most, Canadians now receive their news through Facebook and other social media, news delivery on social media is highly contoured to individual tastes and preferences, and the number of news sources has exploded beyond anything that could have been imagined just a short time ago. Yet news remains the heart and soul of public broadcasting. If the CBC fails on the news front, then one has to question whether there should be a public broadcaster at all.

Over the years, CBC News has changed dramatically at both the national and the local levels, largely mimicking what has happened in the private sector. For one thing, budgets have fallen. According to the CRTC's numbers, Canadian program expenditures in news for both the English- and the French-language CBC fell by close to 10 per cent between 2012 and 2016.[40] Despite the injection of $675 million to the CBC over five years announced by the Trudeau government in 2016 and the extensive hand-wringing about fake news and its negative effects on an informed citizenry, CBC management has been cutting the money allocated to news. It declined in each of the budget years 2017–18 and 2018–19, forcing the cancellation of a daily business news program on News Network and compelling the same news channel to rerun the live top half of each daytime hour in the second half of the hour, during the summer at least. This has also led to the elimination of positions in news bureaus at home and abroad.

In fact, over the past decade, demands on journalists have increased, and the CBC has moved aggressively online, but it says that the overall number of journalists it employs has changed very little. In theory, everyone can now work for radio, television, and online, but the reality is that the numbers dedicated primarily to radio and television to get programs on the air have fallen. At the same time, the ratio of managers to working journalists—one manager for almost every two journalists—seems excessive, but like journalist numbers, it has also changed little. This was borne out by a response from the CBC in March 2019 to an access-to-information request "for a breakdown for the number of CBC employees by year for the last ten years (2009 to 2019) which differentiates between the number of management staff and journalistic staff."[41]

The CBC response is, at best, curious as it seems to exclude producers, associate producers, news writers, editorial assistants, and all the other editorial employees involved in getting programming on the air who do not have a designated technical role.[42]

There is no question that all television and radio newsrooms are smaller now, with fewer reporters (almost no specialist reporters), producers, editors, camera operators, and technical studio staff. This means that news bureaus are simply stretched too thinly to report on at least some stories that should be covered. Stories that are easy and obvious—crime, fires, traffic, weather, court cases—will sometimes get priority because information is provided free of charge by police or fire officials, because the people and resources needed to be in an unfamiliar location or go deeper into a story simply don't exist.

There are fewer correspondents based overseas, which means less emphasis on international coverage on the ground provided by Canadian journalists for Canadians. Much more coverage is now provided by journalists sitting in offices in London, Washington, and even Toronto, editing video from around the world provided by news agencies such as Reuters, Associated Press, or Agence France-Presse and writing scripts based not on being there, but on rewriting wire service stories, which viewers can now read themselves directly for free on numerous

Table 2. Breakdown of Number of CBC Management and Journalistic Staff, 2009–19

Year	Management staff	Journalistic staff
2009	594	1,285
2010	590	1,221
2011	608	1,293
2012	636	1,287
2013	615	1,262
2014	625	1,256
2015	568	1,245
2016	561	1,213
2017	579	1,248
2018	579	1,282
2019	607	1,290

The following classifications are included in "journalistic staff":

ANCHOR-PRODUCER

ANNOUNCER

ANNOUNCER-INTERVIEWER

ANNOUNCER-OPERATOR

ARCHIVAL EDITOR

COLUMNIST

COLUMNIST-RESEARCHER

COMMENTATOR-INTERVIEWER

CORRESPONDENT

HOST

JOURNALIST

JOURNALIST-ANCHOR (SPEC/BLENDED NEWSCAST)

JOURNALIST-PRESENTER

LINEUP EDITOR (NATIONAL)

LINEUP EDITOR (REGIONAL)

LINEUP EDITOR, DIGITAL FORMATS

METEOROLOGIST

NATIONAL REPORTER

NATIONAL REPORTER SPECIALIZED

NEWS EDITOR PRESENTER

PROVINCIAL AFFAIRS REPORTER

REPORTER, DIGITAL FORMATS

REPORTER/EDITOR

RESEARCH JOURNALIST

SENIOR HOST

SENIOR REPORTER

SPECIALIZED REPORTER

TRAFFIC REPORTER

VIDEO JOURNALIST

WEBMASTER COMMENTATOR INTERVIEWER

online sites. The result is a significant deterioration in the quality and depth of the CBC's international coverage. There is simply no adequate replacement for reporters actually being on the ground when something happens, not reporting from thousands of miles away. Nor is parachute reporting, dropping in after a disaster or during a few days, an adequate replacement for reporters living full time in communities around the world and telling Canadians what daily life is like in those locales. At the same time, a change in CBC policy allows freelancers to sign off their stories with "CBC News" rather than "for CBC News," making it appear that the public broadcaster's international reporter contingent is much larger than it actually is.

With shrinking resources, demands on the remaining staff have grown dramatically. Reporters file for multiple media formats and, more recently, are adapting to a digital-first strategy, whereby they file online before anything else. Filing right away and updating as soon as another tidbit is available robs reporters of the time needed to think about how stories should be framed and even whether what they are reporting is actually new or even worth reporting. They are also required to do more in the field, often shooting video or still photos and sending unedited information directly to audiences through Twitter or other social media outlets if they happen to be at the scene of an event, and not, as too frequently happens, covering it from a desk at their office.

Accompanying the decline in numbers of international reporters has been a similar reduction in the number of specialist reporters across the country. It is a damaging trend across all media in Canada. As reporter numbers have been cut and demands grow on those who remain, reporters who once had the luxury of specializing in a specific subject or series of related subjects have been turned into general-assignment reporters. One day they may be covering climate change, the next day a court case, and the day after that a story about social services. There is no time to develop any expertise or go beyond brief interviews and no opportunity to meet people with expertise in any given subject. Research is usually what can be found online through Google in a few minutes before rushing out the door—if, in fact, reporters cover the story from beyond their desks. In many cases, complicated stories simply aren't covered at all as no one has the expertise to sort through what's right and wrong and understand complexities and nuances. Stories about crime, fires and traffic accidents, court cases, surveys and public opinion polls (often delivered by news release from the organization that commissioned them), entertainment, and social media wars are much easier to understand, quicker to turn around, and easier to report. The public broadcaster should be countering this trend to simplistic and largely irrelevant news, but particularly when it comes to local news, the CBC seems to do little that's distinctive.

Over the past two decades, CBC local supper-hour television newscasts have undergone a series of stops and starts. Newscasts were cut, then consolidated into regional newscasts. They were then expanded to 90 minutes (essentially a 30-minute newscast, replayed twice), then returned to an hour-long newscast. Local late-night (11:00 p.m.) newscasts were killed off, then revived, while weekend newscasts suffered the same fate. The only constant in this revolving door of changes was that the number of employees producing the programming kept shrinking, while demands on their time kept growing. These shows are often operated by skeleton staffs, and, not surprisingly, audiences have shrunk to what, in some cases, are embarrassingly low numbers. In some cities, the audiences are a fraction of what they are for private broadcasters. The sidelining of local TV news by the CBC itself has allowed private broadcasters to dominate the local news horizon. Most crucially, they have displaced the CBC as the place to turn to on television when there are local events or an important local news story.

Strangely, CBC local news never produced morning TV news shows (instead choosing to do a national morning program on CBC Newsworld/News Network, available only to cable and satellite subscribers), leaving one of the most coveted spaces in all of news to Global, Citytv, and CTV. CBC placed its bet on its popular drive-time morning and afternoon radio shows, believing, perhaps, that offering early-morning TV news would take away from its bread-and-butter radio audience. It now televises those radio shows in the morning on CBC television stations using fixed cameras in radio studios, not because it is visually compelling but presumably because it is cheap, is easy to do, and contributes to Canadian-content quotas. Everywhere else in the TV world, morning news shows are an essential linchpin linking local stations to their communities and to advertisers, and they are critical for branding. The key is usually to establish a cheery family atmosphere on the set and provide people with the news that they need to "get through the day." For many viewers, if a TV station is not there in the morning, it is simply not there.

To make matters worse, CBC Radio has had its budgets cut as well, despite its strong audience: it has a market share of between 15 and 20 per cent in most major centres, usually giving it the largest single share of the radio audience in those cities and surrounding regions. The cuts have been underway for the best part of a decade. As an example, radio employee numbers were cut by 20.8 per cent between 2009 and 2013 to 1,265.[43] That has led to a significant increase in repeat programming.

The consolidation of radio and television news departments has meant that more and more radio news stories are, in fact, not done for radio but are simply the audio tracks stripped from television stories, with the shortcomings that flow from the radio audience not being able to see the visual elements in the story highlighted in the television script. It goes without saying that writing for radio is different than writing for television, but that has largely been left by the wayside in consolidation. There has also been a significant duplication of content between the lead CBC evening radio news program, *The World at Six*, and its television counterpart, *The National*. Not only do they increasingly cover the same issues, but it is also often done with the same reporter telling a virtually identical story. If you hear it at 6:00 p.m., you don't need to watch it at 10:00 p.m. These are just two manifestations of a greater problem faced by the producers and editors of news and information content at mainstream news organizations.

Until the last decade, it was safe to assume that most of a nightly newscast's audience or a morning newspaper's readership knew little or nothing about the stories contained in that edition or newscast. As mentioned in an earlier chapter, that's no longer the case. With users checking their mobile phones roughly three times an hour,[44] people receive endless bursts of video highlights, Facebook and other social media posts, headlines, prompts, and text messages. The 24-hour news cycle has now been reduced to about an hour. Like freshly baked bread, to use an old analogy, news becomes stale very quickly. News is endlessly replayed, reordered, and replenished, and audience knowledge of the details of individual stories now runs the full spectrum—from some who know everything up to the minute about a story, others who know nothing, and everyone else somewhere in between. This makes it extraordinarily difficult for a producer to line up the order of stories in a newscast or an editor to design a newspaper front page by the old rules of what is most important for the audience to know that they don't know and what they need to know. Today, what's old hat to some is a revelation to others.

It is equally difficult for reporters to determine how to frame and pitch their stories. What should they assume about audience knowledge of the issue, activity or people at the heart of each story? How much do they need to tell listeners and viewers, and what should they conclude is most important for the audience to know? Uncertainty among news editors, producers, and journalists about how to respond has taken its toll. People may listen to or watch the start of the top story in a newscast and say, "I know that," and switch channels. Equally, if a newspaper's front page contains stories from the day before, some readers ask, "Why do I need to buy it?" The result is no surprise. Fewer people are watching television news, and fewer are buying newspapers. This also makes it more difficult to sell online subscriptions behind paywalls.

Audiences for *The National*, the public broadcaster's signature English-language news program, have hemorrhaged considerably since the show's glory days in the 1980s and 1990s. A rough estimate is that *The National* has lost half its audience since 1990, dropping approximately a million viewers. Some of this reflects the fact that CBC "lead-ins"—the programs on immediately before *The National* every night—almost always have a relatively small audience, so that *The National* has to attract viewers from other activities or networks. That's harder to do in the attention economy, with endless ways for audiences to amuse themselves. *CTV National News* has the other challenge—keeping the large lead-in audience that comes from US crime and drama shows that CTV airs at 10:00 p.m., immediately before its national newscast.

Globe and Mail media critic John Doyle has been searing in his assessment of the CBC's flagship news program, claiming that, under Mansbridge, *The National* "declined into a disastrously inane newscast, often an exercise in breathtaking superciliousness."[45] On another occasion during Mansbridge's tenure, he wrote that the show was "sometimes a disgrace, a meandering journey through the mind of a flibbertigibbet who spent the day garnering news bits from a hodgepodge of online sources."[46] As much as we respect Doyle's opinion, there were also many nights when *The National* shone brilliantly—although those nights became fewer and fewer as the years went on.

The new post-Mansbridge *National* launched on 6 November 2017, and from the start, it was clear there were problems. The revamp involved replacing one host with four—two in Toronto, one in Vancouver, and one in Ottawa. (After one season, Andrew Chang in Vancouver and Ian Hanomansing in Toronto swapped places.) The plans also called for making *The National* more than just a newscast. It would be built as and given a multi-platform, distinct identity, separate from the rest of CBC News and its specialty channel, News Network, posting more video to social media, "marking far more online news content with *The National* brand," while highlighting the depth and context of CBC News journalism.[47] Editorial operations would expand, with a stated goal of competing directly against the online presence of newspapers. "There will be a newsletter, podcasts and deeply reported and well-written features known as long reads," the *Globe* reported, with material being published at all times during the day. No longer would the focus be the nightly newscast, thereby reflecting the degree to which people now get information all day long from many different sources. Little of that was revolutionary, but it would require different thinking for all those involved in producing the program. In the end, little of it happened.

The four hosts, who had survived extensive audience testing, had considerable journalistic experience but also played well to the CBC's desire for on-air diversity. One, Andrew Chang in Vancouver, had the added benefit to management of polling particularly well among CTV viewers, the larger newscast audience that the CBC hoped to lure back to its nightly news program. When the hosts appeared on camera, initially frequently presented as a lineup of what looked like baseball cards on the screen, there was no clear delineation of what each host was supposed to do and why. The individual appearance of any of them on screen gave viewers no visual clue of what was coming up next on the program. It did, though, create opportunities for "chit-chat and happy talk" among the anchors about the stories they were presenting—importing a time-consuming staple from local and cable news that, in the past, had been seen as beneath the seriousness of *The National*.

Doyle's critique of the four-anchor format was brutal. He wrote that the show was a "harebrained muddle" as well as "disjointed, surreal and lacking in coherence." To him, the show simply "makes no sense."[48] While Doyle's savaging of *The National* might be seen as harsh because there are moments of poignancy and excellence, the show sometimes comes across as an almost condescending civics lecture or trying, perhaps, too hard to be cool and edgy. The result on too many nights is a disjointed program, even at the best of times, lacking a consistent or coherent voice or sense of purpose.

Part of the difficulty is that visual issues are a significant problem for the program on several levels, not just with the anchors. Video journalism is about showing the audience the story with pictures, not telling it to them with hosts on camera and/or in studios. Yet the mandate of the new *National* is to cast stories forward. Producers decided that the focus should be less about what happened today, which was the traditional role of a nightly newscast, and more about what is going to happen next and what this may mean for the future. The practical problem for television that uses such a forward-looking approach is the lack of pictures about the future to draw upon, to show the audience the story. The default position becomes hosts talking in studio and conducting interviews. That can work in some cases, but if not done carefully and with clear purpose and objectives, it simply risks turning *The National* into a somewhat upscale version of cable news, which at its worst devolves into a room with people sitting around a desk talking or yelling over each other.

In its early days, the new *National* had too much of that sitting around a desk and not enough time spent with hosts outside the studio in the real world, showing stories to its audience. Graphics used to support stories were also surprisingly pedestrian. The program made little use of green-screen technology, regularly employed by the BBC and Global, among others, to place reporters in the middle of the graphic itself, highlighting the key facts and figures that can help explain the story, even without on-the-scene video, or where video is impossible to find, to illustrate the points the reporter wants to make.

This points to a second weakness—the lack of sufficient dedicated video resources for *The National* only, not shared with the News Network or CBC local television stations. For domestic stories, *The National* lost most of its dedicated resources years ago: cutbacks put cameras and editors in local pools, where the technical crews could be working one day on a story for the local supper-hour newscast and the next day on something different for *The National*. Even worse, stories were often put together by reporters melding video shots that had been broadcast earlier in the day with one additional element or an on-camera appearance from the reporter exclusive to *The National*. This explains the growing gap in quality, content, context, and storytelling between domestic and international stories. In the latter case, CBC reporters would travel with their own dedicated cameras to Syria for the civil war, Greece for refugee migrations, or even the US Midwest for trade issues to tell stories that they had crafted on the ground themselves.

The goal of the new *National* was to create an hour of compelling television, different in content and context from everything that audiences might have seen, in whole or in snippets, during the day. On too many nights, however, it has failed to meet that goal. But something else crippled its ability to do so—advertising. The 6 November debut of *The National* had 36 advertisements during the one-hour program, most only 15 seconds long and clustered in three- to five-minute advertising chunks. Some of the same advertisements appeared more than once during the hour, and all of them shattered the flow of the program, offering viewers an easy exit at regular intervals. Research has always shown that many viewers accept that invitation. Had the broadcaster the courage to say that the new *National* would have no advertising, it would have also significantly broadened programming opportunities for those in charge of the show, giving them the flexibility to try different concepts and models, knowing that their program would not be dissected at regular intervals by advertising.

There is no doubt that this would have been costly as it would have meant forgoing revenue, forcing the CBC to make tough decisions about cutting spending in some areas to make an ad-free *National* possible. But it would have been consistent with the CBC's 2016 request in *A Creative Canada,* its pitch to play a new role as part of the federal Heritage department's review of federal government cultural policy, to end advertising and also with its argument that advertising constrained its creativity. It might also have demonstrated to the federal government that the CBC was prepared to take dramatic steps to improve the quality and distinctiveness of its main news program, even if it didn't get more money to do so.

Advertising is a critical handicap. The CBC has been doing everything it can to ram as much advertising into *The National* as possible—to the point where there are now more ads in 30 minutes of the hour-long news program than during the 30-minute *CTV National News* at 11:00 p.m. This ad-cramming has undermined the integrity of *The National* and made it extraordinarily difficult for programmers. Everything has to fit around advertising blocks that break up the rhythm, continuity, emotion, and focus of the program. It forces the hosts to refocus the audience every time the program returns from a commercial break. Commercial breaks are also an opportunity for audiences to see what is on elsewhere or shut off the TV. TV viewing numbers still matter a lot for bringing in that advertising revenue, and *The National*'s audience seems to be slowly vanishing—perhaps due, in part, to how the volume of advertisements undermines the flow and nature of the newscast itself.

As the Canadian Press noted, on a Monday night in January 2017, with Peter Mansbridge as anchor, *The National* drew about 734,000 viewers, falling to 584,00 for the second half of the program. To quote CP directly: "For the debut of the new *National*—now hosted by Ian Hanomansing, Adrienne Arsenault, Rosemary Barton and Andrew Chang— 739,000 viewers were tuned in for the first 30 minutes on CBC, while 601,000 were still watching for the second half. But subsequent nights saw the ratings peak between the high 300,000 to low 600,000 range." As mentioned above, this wasn't helped by the CBC's inability to draw large lead-in audiences from other programs on the CBC's schedule. By contrast, the night of *The National*'s debut, the *CTV National News* had an audience of 841,000, having drawn about a million viewers a night during the preceding week.[49] During the winter of 2017–18, *The National*'s television audience sometimes fell under 300,000, a number far below what some leading podcasters and YouTube pundits command. By the spring of 2019, *The National* had half the audience that watched the *CTV National News,* as television critic Bill Brioux noted on 11 April:

"The *CTV National News* drew 955,000 in overnight estimates. CBC's *The National* did 414,000 and 369,000 over each half hour. An estimated 360,000 watched *Global News at 11*. *CityNews Tonight* informed 62,000."[50]

As we stated at the start of this chapter, news is the very lifeblood of public broadcasting, and *The National* has a prominent place in that as its flagship. If the CBC becomes just another news voice in an increasing cacophony of news choices, if its news presence slowly evaporates, then there is little future for public broadcasting.

Can CBC News Survive?

There are a number of structural issues related to the new digital horizon that go to the heart of what public broadcasting should be and even whether it can survive. One particularly daunting problem is the degree to which online news has become an engineering exercise in which algorithms determine the news that users will receive based on their individual tastes and passions. As described in a previous chapter, Facebook and other media organizations such as *BuzzFeed, Vox*, and the *Huffington Post* increasingly view news as a type of "mood management," whose goal is to keep users cheerful and content. These organizations believe that media analytics can pinpoint the specific desires of individual audiences and match the corresponding stories with ads targeted to their tastes and personalities, including using the types of approaches, graphics, wording, and even colours they like. Ken Auletta has described advertising as having gone from "mad men to math men."[51] Outrage, shock, happy stories, and hot visuals can build traffic, while more serious fare about the economy, the environment, or politics are unlikely to pick up speed. While the ethos in CBC newsrooms can't be compared to that of *BuzzFeed, Slate*, or the *Huffington Post*, large Chartbeat screens tell CBC online reporters whether their stories are trending up or trending down on a minute-to minute basis, a reality that we will discuss at greater length in the next chapter. Critics argue that the very definition of what constitutes news is in the process of being altered. And as Franklin Foer has observed, some news organizations have even changed their news lineups so that "no haters"—that is, no negative stories—get on the air.[52]

The problem for the CBC is that, as discussed above, it is largely "flying blind" where media analytics are concerned, compared to most of the competition, and, more critically, light and breezy, "feel-good" stories are not what public broadcasting is about. Part of the responsibility of a public broadcaster is accountability reporting, which is the kind of reporting that holds public officials and institutions accountable for their actions and provides citizens with a clear-eyed and unvarnished perspective on the issues that will affect their lives. Stories about forest fires, pipelines, refugees crossing the border, the rights of Indigenous people in Canada, emergency rooms, justice, Supreme Court cases, and battles over national unity, among other issues, are integral to the CBC's mandate and must be reported on, even if they put viewers and listeners in a less than happy mood.

In addition, as Cass Sunstein has pointed out, organizations such as the CBC are "general-interest intermediaries," media organizations that, by their very nature, try to appeal to a wide audience.[53] They try to cover a broad range of topics and human experiences, knowing that not all its programs will attract a sizable audience. The problem is that the audience has splintered into narrow, digital cubbyholes based on people's individual passions, interests, and obsessions. Rather than shop at a department store that carries a little bit of everything, they shop at smallish boutiques geared to their interests. Why watch *The National* when there are a myriad of sports, business, crime, entertainment, weather, documentary, and international news channels and websites that can deliver the type of news you want, with points of view that you agree with, when you want it? While some global "general-interest intermediaries" such as the *New York Times* are thriving, many that have a more local reach are in deep trouble.

There are also opportunities opened up by *relational journalism*: the relations that news organizations develop with other news organizations and with their publics. There are certainly avenues for greater co-operation between the public broadcasters and private news organizations, both broadcasters and newspapers, as everyone converges online. In October 2018, for example, the CBC initiated discussions about greater sharing among news organizations in the face of the squeeze they were all experiencing by the collapse of advertising. A precedent had been set in the 1997 federal election, when news organizations faced the prospect of having to pay to put their news crews (in those days, four to five people) on each of five leaders' tours. No news organization could pay for that, so they agreed to pool their coverage. Each of five networks—CBC, CTV, Global, Radio-Canada, and TVA—would cover one leader with a camera, sound recorder, video editor, and producer for the whole six-week campaign. Everything shot was available and fed to all five networks, and the pool crew would shoot individual, on-camera stand-ups and other specific material whenever reporters from each of the five partners were on the tour. It quickly became the standard procedure for all subsequent election campaigns and for foreign trips by the prime minister as well.

The incentive for the CBC to consider doing this more broadly is the same as it is for private news organizations—saving money. Put simply, it is an attempt at "mutually assured survival" in an era where all Canadian news organizations are fighting for their very lives in an environment in which everyone is producing similar content and faces brutal competition wherever they turn. The CBC could, through sharing, save the cost of duplicating coverage of some events deemed less significant but still worthy of being covered.

But there are some large risks for the CBC in following such a strategy. Sharing material reduces the advantage it has in many communities due to its size and its loyal following, on radio in particular. Equally important is whether co-operation with the private sector will lead the CBC to adopt even more of the private sector's approach to news content and presentation than it already has. One can argue that the CBC's local newscasts are already almost indistinguishable from the private sector in content, pacing, and appearance. In addition, the more CBC news and information looks, sounds, and feels like the private sector, the less distinguishable it will be, especially to the transient online audience that already has trouble differentiating the work they see on social media produced by the public broadcaster from that done by others. So much for building the brand through viability on third-party platforms. At what point does the CBC lose its very identity as a media organization?

Another dimension to relational journalism is crowdsourcing. David Fahrenthold of the *Washington Post* won a Pulitzer Prize in 2017 after he asked his readers on social media whether they could offer him leads about Donald Trump's use of donations to his charitable foundation. He in turn let readers see how his story was progressing.[54] CBC journalism has not yet taken advantage of its many opportunities to go "open source" and engage with its audiences in a much deeper way.

Sports and News: The Last Gasps?

The loss of *HNIC* and the relatively low audience numbers registered by *The National*, as well as the CBC's wholesale retreat from local news, go to the very heart of the CBC's legacy and mission. Failure on these two fronts is an indication of deep crisis. As indicated in a previous chapter, the CBC's competitiveness in drama, another genre that has been central to its mission, is in jeopardy because of the revolution brought by Netflix, Amazon Prime, YouTube TV, and other streaming services. Interestingly, the forces that contributed to the crisis in sports and news are the same. Budget cuts and inadequate funding are the smoking guns in both these instances. A lack of funding meant that the CBC could no longer compete for the rights to the NHL and to professional sports in general. The CBC has been knocked out of the game, and there are no indications that the public broadcaster will ever be part of the game again except as a fig leaf for Rogers or another broadcaster.

A lack of funding has also damaged *The National* and other news and current events programs. The pillaging of specialist reporters and foreign bureaus, the thinning ranks of journalists in general, and the lack of technical capacity have left *The National* with much less to offer. While one can add up the show's audiences in different ways across various online platforms to demonstrate that, on aggregate, it still reaches a mass audience, the broadcast may no longer lead or be even part of the national conversation. Nor is it clear that Canadians will automatically turn to CBC News in a crisis to the same extent that they once did. To put it in stark political science terms, *The National* may have lost its "agenda-setting" capacity, its ability to influence the country's great political debates.

The problem in sports and news is that the online world has altered the basic geology of broadcasting. The competition that the CBC faces in both areas is staggering. In sports, every sports broadcaster, league, team, commentator, and even player has a multimedia presence. Moreover, fans have also become broadcasters as they tweet, like, spread, and post hundreds of millions of sports-related items every day. There are also new developments such as fantasy leagues, streaming services, and e-sports specialty channels that have shifted sports into new territories. The question is whether a public broadcaster can afford not to be a part of a cultural experience that is meaningful to so many Canadians and still claim to be a public broadcaster.

The CBC faces the same challenges on the news front. General-interest intermediaries such as the CBC find it much more difficult to capture audiences when so many people receive their news through social media sites that target news to their individual interests, prefer news that entertains them rather than leaves them uncomfortable, and can mine the web for highly specialized news suited to their interests and passions. In terms of international, business, or US news, the CBC has to compete against a myriad of sources, including behemoths such as the *New York Times*, the *Washington Post*, the *Wall Street Journal*, the *Financial Times*, and the *Guardian*, that provide much more complete and, frankly, sophisticated coverage. There are also aggregators such as Reddit, *RealClearPolitics*, the *Huffington Post*, and *National Newswatch*, which provide vast smorgasbords of articles and videos.

The redesigned *National* is also facing a fight for its traditional audience from global media giants that have discovered Canada. The *New York Times* now has three correspondents in Canada, two more than it has had in the past (after covering Canada for some years in the past couple of decades from the United States), with journalists now based in Toronto, Ottawa, and Montreal. Digital subscriptions are now priced in Canadian dollars rather than the traditional US-dollar pricing, and the *Times* now makes a big deal of a weekly newsletter that it sends out, highlighting stories that it has published in the newspaper and online each week about Canada, many of which touch subjects and issues not covered by major Canadian media organizations.

In addition, the BBC in mid-2016 announced plans to hire a video producer, online producer, and social media producer—all based in Toronto—as well as plans to develop "a dedicated version of its North American edition of BBC.com for Canadian audiences"[55] on computers and mobile devices. Both news organizations have their eyes on taking away some of *The National*'s traditional audience.

While the CBC has a natural advantage when it comes to Canadian news and for reliability in the face of an epidemic of disinformation and fake news, it may be slowly squandering these advantages. The larger question is whether CBC News still plays a significant role in sustaining Canadian democracy—acting as a check and watchdog on governmental power, informing citizens about developments in health care and science and as consumers, and providing them with the information that they need to know about their communities and the world. If the CBC is failing in this arena, or is just doing a mediocre job, it is toying with destruction.

Chapter 6

The Trials and Triumphs of the CBC's Online World

In 2009, the CBC decided that it had to stretch its reach across the digital ocean and be seen and heard on as many platforms as possible. In doing so, the public broadcaster registered some notable successes but also encountered significant problems. In this chapter, we describe how the CBC's digital revolution changed the nature of the public broadcaster and brought it into sharp conflict with the private sector.

One problem is that measuring audiences has become more difficult than ever. The loss of audience for *The National* is real, but it also reflects the degree to which, in the attention economy, the traditional way of measuring television audiences is increasingly irrelevant, for many of the reasons noted earlier. The decline in television viewing for news also reflects conscious decisions by CBC management as it has reacted to the new world in which it is operating. CBC News is no longer available just on radio, television, and its cable news channel, CBC News Network. So using *The National* as the canary in the coal mine, as the key indicator of the CBC's success or failure, may no longer be appropriate.

For many years, *The National* has run on CBC News Network at 9:00 p.m. Eastern Time and repeated later each evening for west-coast audiences, while on the main, over-the-air channel, it remains at 10:00 p.m. in time zones across the country. News Network audience numbers may be small in some cases, but they jump considerably during the NHL hockey playoffs, when playoff games bump *The National* from its normal time slots across the country on the main channel. Those who want news and don't care about hockey can get it on News Network. So CBC managers argue that, at the very least, television audience numbers should reflect the combined viewing of *The National* on both channels. But this is hardly a saving grace.

That's just the start. In 2009, when CBC began its digital transformation, management decided that news content should not be confined to CBC television, radio, and its own online sites but should also be available on third-party platforms and apps. Those were still the early days of mobile devices, and CBC chose to focus on smart phones rather than tablets, which has turned out to be a wise decision. While tablets are popular, smart phones are ubiquitous, used everywhere for everything by almost everyone all day long. Making content easily viewable on phones using CBC apps was important, but even more important, the CBC believed, were those third-party distributors. As described in an earlier chapter, Facebook remains a dominant social media redistribution platform for CBC News. But there is also YouTube, Twitter, Instagram, and Snapchat, among many other platforms, and more recently, home devices such as Amazon's Alexa; CBC radio news was the default news source for the device when it launched in Canada.

In other words, the CBC's distribution network is diverse and scattered, and its audiences are amorphous and ever shifting. Michael Wolfe has described the increasingly "low value" viewer as "Not people gathered to pay attention. But people moving to and fro, taking a more often random path, seeing little, absorbing less (attention, that is, time on page, measured in fractions of a second)."[1] Matthew Hindman describes the dominance of "extremely casual viewers," whose page views last an average of 26 seconds.[2] The "bounce rate" among visitors to news sites may be even more fleeting.[3] Hindman cautions that news organizations can easily overestimate their audience numbers because drive-by visitors who stay for 20 seconds, or who are passing through on their way to another site, are "rolled into monthly audience numbers."[4] While the overall numbers may be impressive, the number of real readers and viewers may not be.

How should all these different audiences be counted, and should they be rolled into a single number that represents *The National*'s audience and reach? Even if that is done, can an audience drawn from multiple sources, with users who come and go for short periods of time or disappear within dozens of seconds, be compared to those old TV viewing statistics? Those viewers were watching or listening to a CBC station, and they knew that it was a CBC station. The problem of brand annihilation faced by news organizations whose stories are carried on Facebook, Google's Accelerated Mobile Pages, the Apple News app, or Twitter—and thus lost in an endless shuffle of news stories or promotions—didn't exist. For the CBC, brand loyalty, the stickiness of its audience, isn't just numbers on a scorecard; it's the very currency of survival and the primary defence against budget-cutting politicians.

How important are the audience numbers on all the other media where CBC News now appears? As mentioned previously, while the CBC's online presence is impressive when compared to other Canadian news and entertainment sites, it is low in terms of time—the time that Canadians spend on Facebook, Instagram, Twitter, or WhatsApp. For the CBC's purposes, however, the size of its online audience provides evidence that it is reaching a larger audience than just through TV and radio and, most critically, that it is reaching a younger audience. If it were not for its digital presence, the CBC's audience would be increasingly geriatric.

CBC News online encompasses local and national news sites, individual sites for each radio and television program, and separate community sites for each CBC radio and television station as well as a more recent and widespread push into podcasting. There is an inevitable amount of duplication across such sites, but they also provide local, national, and international news, using their own reporters on the scene in some cases and rewriting other copy in other cases as well as using wire services. Surprisingly, in the CBC's online universe, video can often be difficult to find and certainly not prominent in most cases. This runs counter to the trend toward more video and less print that is evident almost everywhere across the web.

A mid-2018 redesign of all CBC news sites resulted in plunging viewership as audiences found it difficult to figure out the geography of the redesigned sites. They were confusing, with stories often repeated in several sections on the same front page, and it was often hard to determine what was actually important. Subsequent fine-tuning of both the online site and its mobile version for smart phones has done little to solve underlying design shortcomings. This is a reminder, perhaps, that local branding matters and that generic, one-size-fits-all news can backfire.

Chartbeat and the Changing Face of CBC News

In terms of its online content, the CBC has become obsessed with Chartbeat. Large screens with Chartbeat graphics are present in most CBC newsrooms (as well as private-sector newsrooms), giving reporters almost instant feedback on how audiences are responding to their stories. Reporters know what types of headlines work, the types of stories that are popular, which topics chase viewers and listeners away, and the kinds of stories they should pitch to their bosses. They also know the kinds of themes and factors that will cause a story to go viral. Moreover, producers need only check their phones to get an up-to-the-minute score on how popular stories are with users. One CBC producer described Chartbeat's effects this way: "For producers Chartbeat is like a VLT or slot machine. You get 'addicted' to the instant gratification of seeing stories do well. Is my story at the top? How many people are into it? How many engaged moments, etc. Also, there's an adrenalin rush for reporters and producers because bosses aren't doling out compliments for good 'craft' stories."[5] Chartbeat numbers then become a central focus for the next day's story meetings and assignments. In some cases, they can highlight the need to follow up a story the next day as a result of audience interest when the news cycle might otherwise move on. The danger, though, for the CBC and, indeed, for the other Canadian news organizations that use it is that curtailing certain stories because they are difficult or painful gives readers a distorted view of reality.

Another CBC journalist commented on the distorted world that the new Chartbeat culture has created:

> Metrics play a big role in what stories get pursued. Producers actually say, "We should do that story. It will kill online…." Online has also increased the amount of churnalism. Online is an insatiable beast that requires a lot of stories. More stories can turn into more clicks. Also, digital wants certain types of stories: cute kittens or critters, crime, technology, data journalism stories that appeal to younger audiences, things with great visuals, viral stories.[6]

The same reporter noted that Chartbeat numbers also affect TV reporters, although indirectly. They get little acknowledgement for their TV work but are praised if their stories receive a large number of hits online.[7]

One producer thought that Chartbeat had a minimal effect. Stories, in her view, were never "ginned up" to make audiences angry or provoke a popular response. Nor had she heard of any producers who did that. She did find, however, that it was a helpful guide to what audiences were interested in.[8] Like any other situation in which metrics appear to supply an answer, it is easy to become obsessed with numbers—the question is whether those numbers lead in the right direction.

The Problem of Digital Overreach

But did the rush to digital happen too quickly? Former CBC director of research Barry Kiefl, a harsh critic of the corporation's digital-first strategy, argues that television and radio are not disappearing nearly as quickly as the CBC's push to digital might suggest. Kiefl dismisses the importance of the CBC's average-moment audience online, calculating it to be about 10,000. He describes CBC's digital presence as a niche rather than a mass medium, adding that the low numbers are despite 20 years of CBC Internet presence and that "an hour can't go by without CBC radio and TV reminding you a dozen times to check out cbc.ca, yet the audience remains miniscule."[9] In contrast, CBC English and French television have an average-moment audience of more than 500,000, while radio has almost 300,000, according to CRTC statistics.

Gregory Taylor, a professor at the University of Calgary, believes that the CBC's policy is mystifying because the television era is far from over.[10] Conventional television still commands the largest audiences, and it remains the principal gathering place for Canadians. Radio also shows a stubborn resiliency. In fact, one can argue that we are now in the TV era *par excellence* as TV merges with film; and great productions such as HBO's *Game of Thrones* and *The Avengers* are prime examples of the enormous resources that the major studios are investing in TV. While the delivery platforms are now more diverse, TV seems to be growing larger rather than smaller.

The *Globe and Mail*'s John Doyle is in the same camp. He contends that broadcast television is not a dying business. The question is how producers approach TV. He pointed out in August 2017 that US networks like NBC had "figured out how to monetize this digital world and is doing quite well" by pushing its programming online for delayed viewing with fewer ads but targeted more directly to specific audiences, for which the networks could charge higher rates.[11] That's the model CBC was starting to figure out in 2018, with its tentative steps into limited audience data collection, retention, and analysis.

It's hard to escape the conclusion that the CBC's strategy of digital first and digital everywhere may be a miscalculation. By trying to be everywhere—on Facebook, Alexa, Twitter, Snapchat, YouTube, etc., it is expending a great deal of effort for what seems, in some instances, to be very little gain. Far better, perhaps, to put talent and resources into producing high-quality news shows.

Distributing content onto third-party sites and applications does increase the overall number of listeners, viewers, and now readers of CBC-produced material. But that audience is fundamentally different than the CBC's traditional radio and television audiences. Some may see CBC programming on specific YouTube channels, view it on CBC Facebook pages for individual programs, or listen to podcasts through a CBC app. In those cases, there is at least the potential that audiences will make the link between what they are listening to or seeing and who produces it, thereby generating some sort of residual affinity. That creates the possibility of building a relationship with the audience, even if it is weaker than with TV and radio audiences. This is much less likely, however, with those who see CBC news stories passed around on Twitter, Facebook, or other social media. As discussed previously, there it is more likely to appear as just another story amid the attention economy's endless cacophony, where the link between content and media organization is tenuous at best, particularly for younger audiences.

The CBC faces the age-old dilemma confronted by any business—can it afford to pass up what looks like the next big thing? It has decided that it can't. It has targeted Facebook Live and Twitter as part of its news strategy. Unfortunately, the results are dismal. On Twitter, *The National* is viewed by a couple of hundred people most evenings, and Facebook Live events usually register less than a thousand viewers. *The National* on Facebook often has 500 to 600 people watching. But Facebook audiences for breaking news events can be larger. For example, more than 11,000 were watching CBC on Facebook cover the mass funeral for 16 Humboldt, Saskatchewan, junior hockey team members and staff killed in a bus-truck collision in rural Saskatchewan in the spring of 2018. That's a more viable audience than the small numbers that watch other programs, but it was also half the 24,000 watching the funeral online on Global Television's site. As another example, the mid-August 2018 funerals for two police officers shot in Fredericton drew slightly more than 2,000 viewers on the CBC News Facebook page on a summer Saturday. This was double the number watching it on the Global News Facebook page. Interestingly, the CTV News Facebook page didn't show the event, instead redirecting audiences to the main CTV News online site for live coverage.

These are only a couple of examples, but it may suggest that audiences go to the same source online that they choose when watching television. The lesson may be that people no longer automatically turn to the CBC for important news events. The CBC's weakness in the digital world may reflect its larger fall from grace as a news organization. If online viewing follows that trend more generally, the CBC's push to digital as a way to find new audiences faces some strong headwinds.

Then there are timing issues that come with trying to keep pace with the ebbs and flows of the attention economy, particularly when chasing millennial audiences. For example, in the first half of 2018, a small CBC team was developing content specifically for Snapchat, just as Snapchat appeared to be fading in popularity. Catching a wave as it heads to shore is always a tricky business. The risk is that the public broadcaster will be swept back when the wave recedes and be at the mercy of the next breaker.

There appear to be two exceptions to the challenges of finding and keeping a digital audience: YouTube and podcasting. YouTube appears to be a much more successful venture for CBC. Viewers of programs such as *The Fifth Estate* can watch individual stories on the channel, and the online audience is much larger—in the tens of thousands—than for other CBC programs on YouTube. Some episodes, such as its 2018 investigation of an alleged serial killer in Toronto's gay village, had more than 150,000 views by mid-August that year. *Marketplace*, the CBC consumer show, is even more popular online, with 835,000 subscribers to its YouTube channel and, for example, more than a million views of a program using hidden cameras to explore nursing home violence. Most *Marketplace* episodes have between 30,000 and 100,000 views—audiences that dwarf the handful watching on Facebook Live and Twitter. By contrast, *The National*'s items on the CBC News YouTube channel generate a couple of thousand views. These are minuscule audiences in a country of 37 million people, far fewer than many garage-video makers attract on their personal YouTube channels or that ideological warriors such as Joe Rogan, Jordan Peterson, Steven Crowder, Ben Shapiro, or Eric Weinstein capture, even on their worst days.

But is the CBC's relative success with some of its news and current affairs programming on YouTube a reflection of the broader success of its digital strategy? Are these CBC viewers or dedicated YouTube watchers who stumble upon or drop in to see individual CBC program episodes? If the CBC disappeared, would that YouTube audience be upset, or even notice?

It's a question the CBC has difficulty answering. First, it is YouTube and its parent company, Google (Alphabet), not the CBC, that is collecting all the viewer data from CBC programs shown on YouTube channels. What are the demographics of the audience? Are they the same as for TV—in other words, older—or is YouTube capturing the millennials whom the CBC wants? If so, in what numbers? What is the gender of the viewers, and where do they live? What else do they watch on YouTube? How often do they come back to the CBC channels? Does YouTube suggest other CBC stories that viewers may like based on their past viewing habits? Then there is the question that YouTube and Google won't answer—are all the supposed views actually humans watching, or are they bots? More than a few businesses bolster their numbers by using the services of so-called click farms to buy clicks on their sites. All that data goes to YouTube. It isn't the CBC's property, just as it isn't with Facebook, Twitter, or Instagram viewing of CBC programming. Yet answering these questions is vital for digital success.

This situation highlights a related problem that the CBC faces. As mentioned in a previous chapter, even if it could collect that audience data, it's constrained by its status as a public broadcaster. It's a Crown corporation, so the federal privacy commissioner has a role in overseeing whatever the CBC does in collecting and retaining data. So far, the broadcaster has moved gingerly into data collection about audiences, primarily to better target advertising to online viewers. Even there, it is forced to be cautious compared to the private sector. How much data should a government-owned corporation collect about its readers, listeners, and viewers, who should it share it with, and how should it respond when the inevitable political controversy erupts about the CBC collecting and retaining audience data? To date, no one is asking such questions, and so neither the CBC nor the federal government has to answer them.

But if the public broadcaster is not collecting data about its audiences as extensively as its digital competitors, how can it build a long-term future? Without this data, it is destined to be a secondary player, or worse, in the online world. Without the ability to customize ads to individual users, it will never be able to compete effectively for advertising. Without the capacity to build a recommender system, it will be unable to compete against Netflix and Amazon, among so many other sites. Without data, it has to rely on the older, hit-and-miss, sift-and-sort methods of script selection rather than adopt the straight-to-series ordering that characterizes Netflix and other streaming services.

Another issue is that in moving into digital, the public broadcaster has encountered stiff opposition from the private sector. Ken Goldstein has estimated the number of employees and dollars allocated each year for CBC's online operations. Until 2016, the CBC had been including online employee and spending data in the conventional television data it supplied to the CRTC, resulting in both overstating its spending on television and not differentiating its spending on digital services. While the CBC still does not release separate data about its online operations, by comparing the television data in 2016 (which included digital) and 2017 (which did not include digital) that it had filed with the CRTC, Goldstein calculates that "the CBC is spending—annually—between $100–$150 million of its Parliamentary Appropriation (taxpayer dollars) on its digital online services; and ... it is possible that the CBC might have many as 750–1,000 employees working on those digital online services, separate from its employee counts for radio and television."[12] Adding the $42.6 million in digital advertising revenue that the CBC reported in 2017–18 to his low estimate of the parliamentary appropriation spent on digital, Goldstein concludes that the CBC may have digital revenue of close to $150 million, compared to the $171 million reported by the whole Postmedia newspaper chain of more than 65 newspapers across Canada in 2018.

From the perspective of private-sector media companies, while they are all shrinking, the CBC is growing, broadening into an online presence that rivals or exceeds newspapers for breadth of coverage, opening what are, in effect, new radio platforms and spreading its content onto more third-party applications. Each of these moves by CBC is controversial in its own way.

For instance, as part of its digital expansion, the CBC has moved into featuring online opinions and hiring opinion columnists. This seems to be the latest attempt to respond to age-old critiques that the broadcaster is "too left wing" in its news coverage. Critics argue that, in an attention economy flooded with commentary and opinion and at the same time starving for facts, there is no reason why the CBC's online sites should be in the opinion business. Stick to the news, and leave opinion for others. Private news organizations also argue that their paywall strategies are being undermined by a public broadcaster that, with the support of taxpayer dollars, is offering its online news at no cost to readers and viewers. The CBC's response is that it has to go online if it wants to be where its audiences are and that there is now little distinction between broadcasting on the airwaves and broadcasting online. If it buries some of its competitors in the process, so much the better.

In a country where opinion polling has consistently found that only about 9 per cent of Canadians are willing to pay for news and information online, the CBC becomes part of the public's rationale for rejecting paywalls on news sites. The argument is, why should I pay when I can get it for free and news in one form or another seems to be everywhere? Or, to put the complaint differently, a public broadcaster subsidized by taxpayers is helping to put newspapers and other news sites out of business by destroying their paywall models. Ironically, perhaps, the no-cost access to CBC news hasn't produced any outpouring of additional public enthusiasm for the public broadcaster. Even more galling for the private-sector news organizations (mostly newspapers) is that CBC is selling ads on those free sites, undermining a second revenue source that private-sector news organizations believe is essential for them to survive.

There is another, more esoteric, concern for which an answer is difficult to quantify. With news organizations shutting down across the country, often in smaller communities, is the willingness or interest of entrepreneurs and those with ideas for new online news start-ups being dampened because of the advantages enjoyed by the CBC? In other words, is the CBC getting in the way? It has infinitely more reporting and production capacity and access to more money than a local media start-up could hope to match. Related to that is the likelihood that if a local online news outlet proves successful and develops an audience, the public broadcaster may take over what they have carefully built, that the CBC will, in effect, scoop up the winnings. That argument could be made about CBC digital and radio expansions into Hamilton, Kitchener, and London, Ontario, as well as Kelowna, British Columbia, over the past few years. Supporters of public broadcasting argue that there are audiences in those communities that need to be served and that this is precisely what a public broadcaster should be doing. Opponents of public broadcasting say, "Let the market do it." That's one of the issues a new broadcasting act, when it emerges, perhaps in 2020 or later, will have to address.

Left on its own, CBC management, like any corporation, will pursue all viable opportunities for growth and do what it can to disrupt the competition. There remains, though, a solid argument to be made that the size and financial resources of the CBC should not discourage investors from supporting individuals or organizations with new ideas for local media. As our suggestions in the final chapter will highlight, we believe the time has come for the public broadcaster to replace competition with co-operation, particularly with news innovators, creating new models to replace newspapers that have been closing in smaller communities.

The Podcasting Content Bubble

One area that has opened up new possibilities is podcasting. While podcasts are growing in popularity because of the intimacy created by wearing headphones and earbuds, and because of the charm of personal storytelling, with well over 750,000 individual podcasts globally and over 525,000 active shows, a "content bubble" has emerged.[13] Moreover, the average user listens to only seven podcasts per week, so audience attention is limited. Adding to the competition are subscription sites such as Luminary, which assemble and package podcasts and air them without advertising. Canadian podcasts face the dominating presence of American podcasts, such as *This American Life*, *Freakonomics*, *S-Town*, *The Joe Rogan Experience*, *The Daily*, and *Serial*, among others. In fact, in 2016, there was only one Canadian podcast in the top ten on the Canadian iTunes chart.[14] Most Canadian podcasts struggle to survive on meagre advertising and tiny audiences, a situation that sounds a lot like the fate of other types of programming on Canadian television. Arguably, Canadian podcasts simply can't attract enough of a subscriber base to keep going.

A deeper probe undertaken by Brad Clark and Archie McLean identified a top-ten list of Canadian-produced podcasts in 2016–17.[15] In this sample, seven out of ten were CBC podcasts. At least two original series—*Missing & Murdered* (the episode "Finding Cleo" is about a young Cree girl who vanished in the early 1970s, and "Who Killed Alberta Williams?" concerns a missing Indigenous woman whose body was found along the infamous Highway of Tears in British Columbia in 1989) and the cold-case murder series, *Somebody Knows Something* (the first episode of which focused on the disappearance of Sheryl Sheppard in 1999)—are especially noteworthy. As Clark and McLean note, they are expertly produced, reflect the best documentary traditions of public broadcasting, and are serialized stories. Many CBC podcasts are still refashioned radio shows cut in a different format.[16]

Clark and McLean argue that podcasting is one arena in which public broadcasters have a distinct advantage. They have the journalistic chops, staffing, and resources needed to sustain shows in the face of small audiences and have had the ability to attract subscribers. In 2019, the CBC started to exploit this advantage with *Front Burner,* a daily podcast on political issues, which emulated the success the *New York Times* has had with its political podcast, *The Daily.*

While the CBC's online surge is necessary and inevitable and has brought some successes, it has also raised critical questions about the future of public broadcasting. Does the public broadcaster have to be on every platform, even on those where it can barely register an audience heartbeat? Has Chartbeat distorted the news-gathering process and made the public broadcaster seem more like *BuzzFeed* and the *Daily Beast* than the BBC? Is the CBC endangering the paywall models that Canadian newspapers, in particular, need to survive, if they can survive at all, as they transition from print to only online operations? And how can the public broadcaster hope to compete against news organizations and platforms that use data to target users with the news they want and the ads they like?

The CBC's fundamental challenge is that much of the power in the digital world is already locked in. Instagram, *BuzzFeed*, Netflix, Disney, WhatsApp, and Facebook are difficult to displace, and the CBC can survive only if it can get young users to flip to its sites. This means producing information about food and recipes that can compete against *BuzzFeed*'s Tasty, providing health information that can rival news from WebMD or the Mayo Clinic, trying to outdo TMZ for celebrity gossip, and taking young gamers away from Amazon's Twitch. All this is unlikely, if not impossible. Again, it's only in news about Canada and its cities and provinces that the CBC enjoys comparative advantage. If it doesn't move to reinforce this advantage, to own these spaces, then others will.

There is time to make the changes that are needed, but not much time. Parts of the digital transformation may be working, but some clearly aren't. Tough decisions are required, and—most crucially—refocusing to concentrate resources and attention on what CBC does and can do better than anyone else is essential. It's time to discard the rest.

Chapter 7

More Dashed Hopes

On the drawing board, the CBC's 2017 annual public meeting may have seemed like a good idea. In reality, it became the perfect metaphor for the corporation's plight—as a battered relic out of step with the unpredictable attention economy. On an unexpectedly humid, 30-degree day in late September at the end of an exceptionally cool, wet Ottawa summer, the public broadcaster was handing out red winter scarves with CBC logos to students already sweating on a hot day. The students barely broke stride on their way through the lobby of the University of Ottawa's social sciences building.

Not stopping as they headed home or off to class, the students wound their way around a makeshift television studio squeezed into most of the building's foyer for the CBC's annual public meeting.

It was billed as "No Filters: A Conversation about Credibility, Media and the Future of Public Broadcasting," but it was really a curious, late-afternoon, online broadcast. It wasn't a corporate annual meeting and not really a town hall discussion either, although it was live, before an audience of about 200 people sitting at round, cafe-style tables. Some of the audience appeared to be the general public (the CBC had promoted the event online in advance), some were clearly CBC employees, and there were a few students too. They were the ones in shorts and T-shirts.

Then CBC President Hubert Lacroix, in what was supposed to be the final weeks of his term, offered the audience a largely self-congratulatory update on the corporation's ongoing transformation to the digital world. He opened the event with what had become almost a boilerplate element of all his recent appearances and speeches. He highlighted the extent of CBC's digital transformation, the size of its social media audience, the more than 200 million CBC podcast downloads since 2016, and the number of employees who had been retrained in the last couple of years. Not surprisingly, he did not mention those positions and jobs eliminated due to cutbacks in the broadcaster's parliamentary appropriation under the Conservative government before the fall of 2015.

One-third of the way through the hour-long event, six CBC/Radio-Canada journalists took to the stage to answer questions from the audience in both languages. Without providing any further details or context about the state of the CBC, the questioning from the audience and the responses from the journalists rambled from topic to topic—in part, because the journalists on the stage weren't the ones to whom questions about accountability needed to be addressed. They were on-air talent, not editorial or corporate management. A CBC communications person hovered at the media table, occasionally speaking with the only reporter there to cover the event. That reporter was from Canadian Press, and he was wasn't sure whether there was anything worth reporting. He decided that there wasn't a story.

In a world in which the CBC is desperately trying to connect with its audiences, and especially younger audiences, the public meeting at the University of Ottawa seemed contrived and gimmicky. As one CBC employee who attended the event said later, "It felt like something organized primarily to check off a box on a list of things the corporation had to do in its role as a public broadcaster. Hold a public meeting nominally focused around a journalistic issue—check! Done that for another year."

The CBC's Blueprint for Change

It wouldn't be surprising, though, if the attention of senior management that day was elsewhere. The public meeting was held two days before Heritage Minister Mélanie Joly released *Creative Canada: A Vision for Canada's Creative Industries*, her much anticipated and discussed blueprint for overhauling government support for all forms of culture, including the CBC. What was expected and had been promoted as such by the minister for months was a long-overdue update of cultural policy for the digital present and future—a 21st-century makeover.

The CBC had a lot riding on the report. In the late fall of 2016, it had released *A Creative Canada: Strengthening Canadian Culture in a Digital World*, which was at once its own vision for the future and its recommendations for the minister's cultural-policy review process. Readers may be confused by the fact that Joly's report and the CBC's vision document have similar titles. The CBC document described the role that the public broadcaster believed it should play as the spearhead and centrepiece of the country's cultural policies. The CBC would be back at the helm as the creative hub and main marketer of Canadian culture. In many ways, it read as if CBC management would like nothing more than to turn back the clock, reinstating everything it used to do during the preceding almost four decades that had fallen by the wayside. The document, however, was more than just a trip into a nostalgic past, a call to bring back the good old days.

There was an audacious request for an additional approximately $400 million per year in its parliamentary appropriation, which would allow the CBC to remove all advertising from television and from its growing online presence. It is not that the idea of getting out of advertising was bold or new. Advocates for public broadcasting had been calling for it for years, but perhaps it was finally time to make a virtue out of what had become a difficult and all too obvious necessity.

The loss of NHL hockey rights had reopened the conversation. With advertising disappearing, was it time to ban all ads on CBC television to match what had occurred decades ago with radio? The corporation's decision to revive advertising in 2014 on its Radio 2 service (now CBC Music) had been a dismal failure as the initiative raised only $1.2 million. The CRTC had shut down the experiment in August 2016, when it ruled that the CBC had failed to demonstrate that it was using the advertising proceeds to invest in radio. In addition, with Justin Trudeau's Liberal government providing the broadcaster with an extra $675 million over five years in its spring 2016 budget, the CRTC had concluded that the CBC no longer needed Radio 2 advertising revenue.

In fact, in the first year without hockey, net advertising revenue had become small potatoes for the CBC. Barry Kiefl, president of Canadian Media Research Inc., noted, "In the year ending August 2015, CBC English TV ad revenue fell off a cliff and was barely $100 million, well under 20 per cent of TV revenues. Funding from taxpayers is now four times greater than ad revenues.... To make matters worse the cost of selling ads and promotion on CBC TV represented $62 million last year. Administration costs, some of which are for sales, were another $78 million. After deducting these overheads, there is no meaningful profit to be made from advertising, which begs the question: why is CBC still in the advertising game?" Kiefl also observed, "Ad revenues from CBC Internet digital services, such as CBC.ca, are apparently so embarrassingly low that they are folded into the revenues of TV/Radio or buried." The numbers were just as dismal on the French-language side.[1]

In testimony before the House of Commons Standing Committee on Canadian Heritage in October 2016, Hubert Lacroix tried to put a positive spin on the loss of hockey by pointing out that the broadcaster did not make any money on the six-year NHL contract that ended in 2014. The image of *HNIC* as a money machine was a myth. Lacroix argued that its value lay in the fact that "it was an important locomotive in the context of what Canadians wanted to see" and that it allowed the CBC to bundle offers to advertisers so that hockey was part of the package that they would receive if they advertised on other shows.[2] Now, of course, Rogers brings in that hockey advertising revenue, and while the CBC can promote its own programs on the Rogers broadcasts, it is unable to bundle advertising on those shows with hockey to make a more attractive package for ad buyers.

Ending advertising could also lower the temperature of a simmering dispute with private-sector news organizations. The Heritage committee, in that same examination in which Lacroix had testified, heard complaints from the private sector, asking why it had to compete with the publicly funded broadcaster for online advertising revenue at a time when advertising was drying up for newspapers, magazines, and conventional TV stations. For those organizations, online advertising held out at least some prospect for revenue growth that would keep them profitable and alive. They objected to fighting the CBC with its 27 television stations and 41 regional websites (plus 88 radio stations that were ad-free) for what was fast becoming an advertising wasteland. As discussed previously, Google and Facebook account for the lion's share of online advertising, leaving very little to sustain other media.

Lacroix dismissed that complaint of private-sector critics when he appeared before the committee that November:

> Last year we earned $600 million in self-generated revenue of which $253.2 million was advertising revenue. Just ten per cent of that advertising revenue, $25 million dollars, came from all digital advertising across CBC/Radio-Canada. To put that in context, total digital advertising in Canada generates over $4.6 billion a year, three-quarters of which goes to Google, Facebook and Yellow Pages. It is difficult to believe, as some media have suggested, that if only CBC/Radio-Canada, was prevented from earning $25 million, their problems would be solved.[3]

Lacroix was right, of course, although digital advertising grew to $42.6 million for the year ending 31 March 2018.[4] Some critics, however, thought that it was tone-deaf for Lacroix to be goading the commercial broadcasters when the CBC had just received a jolt of new cash from the government. Even small morsels of advertising leftovers had begun to look good.

The pitch for an ad-free CBC captured the headlines, but there was more than that in the proposal under the headings of digital innovation, contributing to a shared national consciousness and identity, creating quality Canadian content, and promoting Canada to the world.[5] All these wonders, the CBC proposed, would happen through "a cohesive cultural investment strategy" led by a Canadian cultural industries council based on the Creative Britain model, as would regulating existing broadcasters and new media entries on an equal footing to support Canadian content. The Creative Britain model in the United Kingdom was largely the invention of Chris Smith, who was an influential cabinet minister in the Labour government in the late 1990s. He had argued that culture was one of the central spokes of the British economy, one that permeated all aspects of civic life. It also had to be at the very heart of the government's political agenda, harnessing both public and private funding, and include grassroots decision-making. The CBC's reference to the British model seemed little more than a throwaway phrase, with little effort to map out what it would actually look like or how it would work.

The proposal that was at the heart of the CBC's *A Creative Canada* was for per capita funding for the CBC to rise to $46, up from $34 in 2017. Funding would be linked to a five-year licence renewal and be indexed to inflation so that it would not be chipped away at over time. Most critically, it would not be subject to the vagaries of elections and government budgets. After all, as the document immodestly stated about the CBC, "Our value to the country goes well beyond informing and entertaining; we're at the very heart of Canada's cultural ecosystem."

While long-standing institutions such as the National Film Board and the Canada Council for the Arts might object to the public broadcaster's view of itself as being the most glittering star in the Canadian cultural cosmos, given that this was the CBC's big pitch, its big push, to guarantee its future, modesty was not a virtue. The goal was to persuade the government that some direction and focus was needed—something governments had been unwilling to commit to for decades.

The document further suggested that the CBC would lead Canada's charge into the digital universe. The English-language service would replicate Radio-Canada's experience with its streaming services Tou.TV and Tou.TV Extra, which it did when it created Gem in 2018. In addition, the CBC stated that it would continue to expand its distribution of digital content using YouTube and Facebook as additional platforms for entertainment and news programming, including for its flagship newscasts *The National* and *Le Téléjournal*. The need to make these last moves had become so obvious and elemental that even stating them as goals seemed unnecessary—except perhaps to satisfy governments that the CBC "got" the new digital world.

The CBC also proposed that it step away from the commitment that it had undertaken under the Broadcasting Act of 1991 to promote national consciousness and identity. This commitment had replaced an obligation placed on the CBC in the Broadcasting Act of 1968 to promote national unity. In its place was a pledge that the CBC hoped would capture what it saw as the new spirit of the times—and, not unsurprisingly, one of the main themes of the Trudeau government—"to be an enabler of social cohesion, giving Canadians unparalleled access to information and programming that reflects a diversity of voices and perspectives."[6] It would do so through partnerships and by increasing Canadian content online. Interestingly, this new wording might have had the effect of removing the CBC from the line of fire of ego-driven prime ministers who believed that their interest was the national interest and their perception that CBC had an obligation to report on the government in a positive light. Presumably, the CBC would have a further guarantee that it did not have to dance to anyone's tune but its own.

A main point in *A Creative Canada* was that the CBC would end all advertising. This would allow it to reinvent itself according to different rules. As the document boasts,

> Our focus would be more firmly on the needs of citizens, creators and our industry partners without the constant preoccupation of monetizing each of our initiatives.... We would focus less on commercial return and more on cultural impact, exploring more ways to help Canadian content and creators thrive and grow. We would be able to commission programming that takes risks and has the time to find an audience without being overly driven by the need to deliver immediate success.[7]

The *Globe and Mail*'s Kate Taylor supported the concept, suggesting that "in an increasingly scattered but ever more Internet-dependent and globalized media environment, the country needs a public producer, curator and distributor to craft a powerful Brand Canada across all platforms, offering not only news, public affairs and documentaries but also fiction, variety and arts programming. It needs an iconic institution to nurture and lead the cultural industries, a rallying point for Canadian creativity." Her enthusiasm was tempered by the concern— "leaving aside, for the moment, the political realism of the request"— that "the danger of the CBC proposal, however, is that the country and the government will seize on the idea while fudging on the money needed to make it work."[8]

In its news coverage of the CBC proposal, the *Globe and Mail* drew attention to those political realities, noting that the request required the parliamentary appropriation to rise by 34 per cent, to $1.633 billion from $1.215 billion in 2017–18:[9] an additional $418 million annually. That would offset the $253 million the CBC said that it currently received in advertising after deducting the $40 million it spent on the business of selling those ads, $105 million for content that would take the place of the ads on air, and $100 million for "new investments to face consumer and technology disruption."[10]

Supporters of an ad-free broadcaster, including a group of former employees under the banner Public Broadcasting for Canada in the 21st Century, were, in the autumn of 2016, the latest of many groups to put their oars in the water, pressing for an end to ads. But a member of the group, Jeffrey Dvorkin, a former managing editor of CBC Radio News and now a journalism professor at the University of Toronto, pinpointed what he saw as a weak link in the CBC's proposal. He expressed a common sentiment after the document was released, arguing that the CBC didn't need more money to put an end to advertising on television and online. He bluntly noted, "They need to figure out what their priorities are, make some hard choices, learn to live within their budgets and then provide the programming that serves Canadians as citizens, not just as consumers."[11]

Others were even harsher. The *Globe and Mail's* John Doyle used a May 2017 column on the CBC, announcing its fall television season, to challenge the themes that underpinned *A Creative Canada*:

> The CBC's own perception of itself is that in a chaotic, shifting media landscape, the CBC is reliable, trusted and more Canadian than anything or anybody in the country.
>
> Sitting as it does, awkwardly at the intersection of commerce, culture and mandated nation-building role, it is dangerous for the public broadcaster to think of itself as defining the country and its culture.... it does not define Canada or the culture. Nobody does that. The country is too mercurial, shifting and happily shapeless—it's an idea with many histories and cultures.
>
> When we hear any organization claim to be defining a country and culture, we should be nervous and suspicious. Some dubious kind of pseudo-patriotism is behind it all. Canadians consume a lot more than what the CBC offers across its platforms and they are no less Canadian for doing so.[12]

Doyle's blazing critique struck at the core of much of how the CBC has responded to the attention economy. As a public broadcaster, the corporation has always wanted to do more but has never been prepared or allowed to stop doing anything it was already doing so that it could find the money within its existing budget to launch something new. *Creative Canada* captured that expansive spirit perfectly. There is more that we can and must do for Canada and Canadian culture, it proclaimed, but we shouldn't have to stop doing anything we are doing now to do it. Perhaps senior managers at the corporation felt emboldened about proposing that it drop advertising in exchange for a larger parliamentary appropriation by the arrival of what they hoped would be a sympathetic Trudeau government. As we have pointed out in this book, with the loss of revenue from *HNIC* and with only pitiful advertising crumbs left on the table by Google and Facebook, the CBC was making a virtue out of what promised to be a bleak advertising future.

A suggestion from controversial former CBC English-language vice-president Richard Stursberg that Parliament should draft a charter for the CBC, modelled on one that established a vision and direction for the BBC in the United Kingdom, generated little response either within or outside government. He had argued that such a charter "must establish a vision for the CBC, so that it [the federal government] can make sensible decisions about how much new money to provide. It should not give the corporation more money until there is a clear understanding about the direction it needs to pursue."[13] That understanding would prove to be more elusive than ever.

Netflix's Canada Policy

If continued support was the CBC's expectation from the cultural-policy review process, the details of Mélanie Joly's September 2017 announcement must have come as a shock. Joly, a young and highly regarded lawyer who had gone to Oxford and run for mayor of Montreal (finishing a strong second), was a rising star in the Liberal Party. The Heritage portfolio was just the first stop in what many, including her, expected to be a climb to the top, or at least near the top, of Canadian public life. Moreover, *A Creative Canada* was to be her signpost—a demonstration that she had the imagination and determination needed to bring about important changes. Unfortunately, her policies proved to be a disappointment and raised widespread concerns about her ability to manoeuvre amid the big fish.

Strangely, the CBC's recommendations, its blueprint for the 21st century, were completely ignored. The government accepted nothing from the CBC's *A Creative Canada* beyond offering vague support for the creation of a Canadian cultural industries council. There was no endorsement of the CBC's view that it played a pre-eminent role in the development and promotion of Canadian culture, no defined place for it in the global marketing of Canadian culture, and no reference to the idea of an ad-free CBC. Even worse, from the public broadcaster's point of view, the fulcrum of the cultural-policy review, proclaimed triumphantly by Joly, was a vague commitment by Netflix to spend $500 million over five years in Canada (including $25 million on French-language productions and to establish a production centre in Montreal) to produce programming that presumably it would distribute globally. There was no indication that the programming would be produced solely in Canada or that it would have Canadian themes. Nor would Netflix be required to contribute any portion of its revenue from Canadian subscribers into a fund such as the Canada Media Fund to develop Canadian content, unlike the funding requirements placed on its Canadian broadcasting competitors.

In exchange, although this was unspoken, Netflix would pay no taxes and would not be subject to Canadian content regulations, and Canadian subscribers would not have to pay HST or GST on their subscriptions. Curiously, Netflix executives did not show up for the minister's announcement, no new shows were unveiled, and there were no details about spending commitments. The deal seemed to be largely a one-way street, with the government allowing Netflix to operate unencumbered by regulations or taxes while getting only amorphous and perhaps imaginary promises in return. The policy document read more like Netflix's Canada policy—which was to do as little as possible while avoiding any responsibilities—than Canada's Netflix policy.

Interestingly, Joly had undertaken a tour of the FAANGs in late April 2017, visiting the head offices of Amazon, Facebook, Netflix, and Google. In one of those "hitting a brick wall" moments, she confessed that "her pitch for cultural diversity was 'surprising' to executives." Joly told reporters that her approach to the giant media platforms was not all carrot: "'There's always a stick,' she said, 'and they know it.'"[14] When push came to shove, however, the stick was nowhere in sight.

Joly soon faced a tidal wave of criticism. Some worried that the policy would create a two-tiered broadcasting system—one for domestic producers, which would still have to pay taxes and contribute to the Canada Media Fund, and one for foreign streaming platforms, which would have few, if any, obligations to Canada. Reaction in Quebec was especially harsh as the $25 million that Netflix designated for French-language productions was seen as little more than a token gesture, at best a drop in the bucket. There was also the suspicion that

Netflix would invest only in headline-catching shows such as *Anne*, based on *Anne of Green Gables*, which could be promoted around the world. In other words, Canada might provide the scenery, a few jobs, and little else.

Some argued that the policy merely publicized commitments that Netflix was already making. The *Globe and Mail*'s John Doyle described the announcement as "a sweet deal for Netflix" and noted:

> Besides, as Netflix will tell you, it has been supporting Canadian TV for years. It spent millions to revive *Trailer Park Boys*. It spent money to acquire the outside-Canada right[s] to Discovery Channel's *Frontier*, CBC's *Anne* and *Alias Grace*, the Showcase series *Travelers*, and other series. It has been behind *Orphan Black*, which in some markets has been presented as "A Netflix original."
>
> In fact, in Netflix's submission to [Minister] Joly's consultation on a reset for Canadian culture in the digital age … the company states, "In 2016 alone, we've commissioned hundreds of millions of dollars of original programming produced in Canada."[15]

Ian Morrison, spokesperson for the lobby group Friends of Canadian Broadcasting, also criticized *Creative Canada*. He accused the Trudeau government of "'punting' key questions down the road, possibly until after the next election." His view was that "there was an absence of protein in this document and they wanted the Netflix stuff to try to hide that."[16]

Richard Stursberg, a former head of English-language services at the CBC, thinks that the federal government made a key error in its Netflix policy. He believes that a dose of tough love would have been far more effective in levelling the playing field and ensuring much more Canadian content programming. He argues that the FAANGs, including Netflix, should collect the GST and the HST on their ad sales in Canada and that they be treated as foreign entities under the provisions of Bill C-58, which penalizes Canadian companies for advertising in foreign media. In addition, Canada should adopt the European Union's policy of forcing Netflix to devote 30 per cent of its programming to European content and a proportional amount to that of individual countries. He also notes that countries such as France, Norway, Australia, South Korea, and Japan have already imposed a Netflix tax. The problem, of course, is that if streaming services such as Netflix and Amazon walked away from Canada, there would likely be wholesale rebellion by Canadian consumers, who have already made binge-watching from a smorgasbord of choices at relatively cheap prices a way of life. Canadian political leaders might be reluctant to tamper with this hornet's nest.[17]

Nevertheless, Justin Trudeau during the frenzy of the 2019 federal election campaign announced that his government would tax foreign media platforms as a way to pay at least in part for his election promises. He did not provide details.

Also worrying was that while Netflix is currently the largest streaming platform, other Internet broadcasters such as Amazon Prime, YouTube TV, Disney Plus, Apple TV+, and HBO Plus, among a host of other large industry players, are quickly catching up. By betting on only a single horse, the federal government might have miscalculated in choosing the wrong winner in the global-streaming sweepstakes. Most observers predict that when the sifting and sorting is through, there will be only three or four giant corporations that have the financial might, global partnerships, and quality programming needed to dominate the streaming industry. Netflix, Amazon, Disney, Apple, and Google (through YouTube TV) are all in the running. A handful of other players, including CBS, Showtime, and HBO, to name only a few, are also likely to find large audiences for their streaming services. As an example of other potential players not included in the announcement, in 2017 Hulu produced and aired a much acclaimed, ten-episode adaptation of Margaret Atwood's *The Handmaid's Tale*.

Amanda Lotz, perhaps the foremost expert on how the Internet has revolutionized TV, suggests that viewers in the future are likely to buy one or more packages of services—a Netflix package, a YouTube TV package, a Disney package, presumably a Bell Media package, a Rogers package, an Amazon package, etc. that bundles a whole phalanx of conventional, cable, streaming, and mobile services.[18] Most consumers will be able to afford to buy only two or three of these media ecosystems. There are two questions: Which two or three will win the battle for consumer eyeballs? Will any of them be Canadian?

Perhaps the real message behind Joly's announcement is that Canadian producers, including the CBC, have now reached the point where they will have to team up with Netflix and other global streaming platforms to reach Canadian audiences. To put it differently, perhaps the government finally acknowledged what has been obvious for some time: that we have reached the end of a Canadian broadcasting era and that Canadian producers can no longer create shows that penetrate popular culture without the help and exposure of one of the giant streaming platforms. Without a Netflix or Amazon or Disney booster rocket, they are unlikely to reach sizable audiences.

The new cultural policy's most startling omission was the complete absence of any recognition that journalism—the linchpin of Canadian democracy—was in deep crisis. An initial government response on that issue came in the February 2018 federal budget, which included

$50 million over five year to "support local journalism in underserved communities." In addition, the budget indicated that the federal government would explore "new models that enable private giving and philanthropic support for trusted, professional, non-profit journalism and local news."[19] Details were to follow here as well. The Heritage department launched a consultation process to determine, first, which arm's-length organization should be responsible for distributing the $50 million and then to develop the criteria for applicants and recipients. By November 2018, there were still no details on how this scheme would operate, but the Liberal government in its fall economic update added an additional $595 million over the next five years to support newspapers struggling to survive the dramatic and ongoing loss of advertising revenue that had once made up about 80 per cent of newspaper revenue.[20]

More details came in the March 2019 budget. Even then the journalism bailout remained a work in progress: key decisions had yet to be made about the makeup and detailed mandate of an independent panel that would write more specific criteria for the subsidies and presumably determine eligibility as well.

The government allocated $360 million over five years starting on 1 January 2019 for labour subsidies—refundable tax credits of up to $13,750 per employee—for news organizations that provided general-interest news. Broadcast outlets and specialist publications of any sort were excluded from the program. An additional $138 million was set aside for 15 per cent non-refundable tax credits for consumers who purchased digital subscriptions to online news sites; individuals could receive a maximum tax credit of $5 per year for digital subscriptions. Finally, the budget proposed spending $96 million over five years to allow news organizations that wished to do so to restructure themselves as not-for-profits or charities, which could issue charitable-donation tax receipts. The various subsidies would be open only to Canadian-owned and -controlled entities, although this could lead to national-treatment challenges from organizations such as the *New York Times* under Canada's trade agreements such as NAFTA.

As Christopher Waddell, one of the authors of this book, has argued, there were key elements missing in the budget announcement that should be front and centre in any government-subsidy program.[21] First and foremost, there was no indication of what the objective was in spending almost $600 million, nor any requirement for specific action by news-organization recipients, such as making the transition from print to online. Nor was there any sense of whether news organizations had to meet annual performance criteria in order to benefit from the subsidy for all five years. With no details on any of that, there is no way of assessing success and whether the subsidy has achieved its objectives.

Similarly, the consumer-subscription tax credits seem destined overwhelmingly to reward those who are already subscribing to digital publications. Since opinion polls consistently show that about 9 per cent of Canadians are willing to pay for news online, the problem isn't that subscriptions are too expensive. It's that the public sees news everywhere for free and asks why it should pay anything. A 15 per cent tax credit won't change that.

Turning news organizations that once reported 20 to 30 per cent rates of return into not-for-profits may benefit a few entities. But there is no history of philanthropic giving by Canadian charitable foundations or individuals for media ventures or in defence of press freedom or freedom of expression, as there is in the United States. Additionally, the money allocated to support such ventures over five years isn't actually very much given the extent of the need, suggesting that there are limits to what the government can do.

Finally, there was no mention at all in the journalism package of the CBC or the role that the federal government believed its public broadcaster should play in the evolution of the Canadian journalism landscape, similar to the way the CBC's proposals had been ignored in the Heritage department's *Creative Canada*. That omission seemed particularly egregious, considering the extent to which the private sector had complained about competition from the CBC for digital advertising revenue, let alone the argument that could be made that Canadians won't subscribe to online news sites as long as the CBC makes all its news available online at no cost.

While *Creative Canada* stressed the importance of the Canada Media Fund, which finances Canadian productions, Joly made no financial commitments to it, despite the fact that Canadian media organizations were finding it increasingly difficult to keep up their contributions as their bottom lines sank. Such a commitment came in the 2018 budget, however, with the promise of $172 million over five years to offset declining contributions to the fund.[22]

What is remarkable is that $172 million stretched over five years is less than the cost of producing a single leading Hollywood TV series or video game. *Game of Thrones*, for instance, had a $15 million-per-episode budget in its final season. Amazon paid over $200 million to produce a *Lord of the Rings*–type fantasy series, which is years from release and is expected to cost much more by the time it debuts. The barriers to entry to the entertainment big leagues are now well beyond what the Canadian media industry seems able to afford.

A point made previously is that, in the blockbuster culture described so aptly by Harvard Professor Anita Elberse, there is far greater risk in producing TV shows with lower production budgets than with higher ones.[23] Small bets tend to lose money, while big bets tend to garner the most winnings. In a universe where there are 500+ new shows a year, Canadian TV producers that depend on the Canada Media Fund play in a low-stakes game in which the odds of winning are stacked against them.[24]

At the same time, there was not even a whisper about requiring Netflix·or any other streaming platform to contribute to the fund. For the purposes of supporting the Canadian media system, foreign companies literally take the money and run.

The stark reality is that Joly and the Liberal government chose Netflix, not the CBC, to be Canada's cultural merchandiser. Just as was the case with the decision to launch a wave of specialty channels on cable in the 1980s, the CBC was an afterthought. Even more galling was that Netflix then passed on its spending commitment to its Canadian subscribers in higher monthly subscription costs—just like the cable and satellite companies do.

Analyzing the Netflix spending pledge, the Solutions Research Group, a consulting firm, observed that, in 2018, Netflix customers paid between $8.99 and $13.99 month in subscriber fees, and most paid the standard plan of $10.99. That gave Netflix about $778 million in revenue from Canadian subscribers. According to the Solutions Research Group, "Netflix's agreed minimum spending on Canadian programming is closer to 13 per cent of its revenues [in Canada], which is less than the nearly 19 per cent that Canadian TV companies pay into the [Canada Media Fund] system."[25] While the federal government gave Netflix special status, it did not move the yardsticks for Canadian broadcasters, which had to compete against Netflix.

That situation gave a foreign company a distinct pricing advantage in competing against Canadian streaming services such as the CBC's Gem, Bell's Crave TV, and Rogers's Sportsnet NOW as well as domestic online news sites, accessible behind paywalls, where Canadian consumers had to pay GST or HST on their subscriptions. The same discrimination against Canadian media applied to advertising. Those buying advertisements through Google and Facebook were not charged GST or HST when they purchased ads, while those advertising on Canadian online sites with advertising sold by Canadian organizations had to pay the tax, which can be as high as 15 per cent in some provinces. The

2018 and 2019 federal budgets chose not to address either of these obvious inequalities.

At the same time, *Creative Canada* held out the promise that the government would review the 1991 Broadcasting Act in the near future. Earlier statements from the Department of Canadian Heritage at the consultation stage of the process had indicated that eight federal acts would be up for review as well as the mandates for the CBC, the CRTC, and the Canada Council.[26] Later in 2018, the government named a committee of experts to review the Broadcasting Act and make recommendations by the end of January 2020.[27] So in effect, the Trudeau government's entire first term in office would pass without any real action taken on the fundamental, if creaking, structures of Canadian broadcasting and communications policy. Issues and problems such as net neutrality, hacking, privacy protection, the use of data analytics, online advertising, and the production of fake news and disinformation on an industrial scale were barely on anyone's horizon. Personal computers and cellphones were in relative infancy, and social media platforms had yet to arrive when the Broadcasting Act became law; by any conceivable standard, the government's response would be at least a decade too late.

In sum, on virtually every issue surrounding Canada's digital future, the document simply kicked the can, including the "Can" in CanCon (Canadian content), down the road to some future date, leaving decisions perhaps to a different government. In the process, the Liberals potentially surrendered an opportunity to reshape the government–cultural industries relationship and the role of the CBC for decades to come.

Not surprisingly, Joly's star quickly fell. The harsh criticism and bewilderment that her performance received in English-speaking Canada paled in comparison to the roasting she took from the French-language media. Her inability to explain the details of the Netflix arrangement and, more specifically, the lack of any significant commitment for French-language production by Netflix were widely condemned.

Political columnist Chantal Hébert concisely quantified the issue:

About half of all this country's households subscribe to Netflix. Canada happens to be home to the largest number of citizens whose mother tongue is French outside of France. Based on the Canadian content available on Netflix, this country might as well be a unilingual cultural colony of the United States.… A total of two television series appear in Netflix's "binge-worthy Canadian TV dramas" category: *Travelers* and *Between* and that's two more than are listed as binge-worthy in French.[28]

A disastrous appearance on the popular Sunday-night Radio-Canada television talk show *Tout le monde en parle* sent Joly's political fortunes into a tailspin, from which she has arguably never recovered. By the end of 2017, she had become a subject of ridicule in the Quebec media for the robotic manner in which she responded to questions, giving pre-programmed comments that often ignored the subject of the questions. According to one commentator, "Her fall from grace in her home province has been swift and merciless, sped by her maladroit attempts to sell a deal with Netflix that would give the company a free pass from tax and regulation in exchange for an ill-defined Cancon investment of $500-million over five years. The Minister has been roasted and ridiculed to her face on live radio and TV and dismissed by commentators of all stripes as naive and—worst of all—unable even to understand what the fuss is about."[29]

In July 2018, Joly's star finally fell to earth. Montreal MP Pablo Rodriguez replaced her as Heritage minister, while she was shuffled to a lower-profile Cabinet position.

To add insult to injury, in fall 2018 Netflix raised its rates. The price for its basic package went up by a $1 a month, while its standard and premium packages increased by $3 a month. In other words, instead of Canadians taxing Netflix, Netflix taxed Canadians.

No Longer a Political Football?

When it came to the CBC, the government seemed content to throw more money at the corporation and do little else. Hubert Lacroix reached the end of his term as president and CEO in December 2017, and the prime minister did not name a successor until April 2018, when he chose TV and film industry veteran Catherine Tait, who took the helm in July 2018. She would be the CBC's first female president and CEO. To Trudeau's credit, Tait was widely seen as a solid choice. As well, departures from the ten-person, part-time board of directors as members reached the end of their terms meant that, by November 2017, the CBC had only the minimum number of directors required for quorum at meetings. Again to its credit, the Trudeau government created an independent advisory committee to accept applications for the board and make recommendations to the government. Gender equality and merit were the criteria for selection.

While the creation of a neutral advisory committee is a welcome departure from the past, the prime minister still appoints the president and CEO as well as the chair of the board of directors. The fly in the ointment is that, unlike in corporate settings, where presidents and CEOs are chosen by and answerable to their boards and appointed almost entirely because of their experience in and knowledge of their industries, CBC board members have had no say in selecting board chairs and presidents. Moreover, since politics rather than industry experience has been the main criterion for selection, presidents and chairpersons have ranged from eminent journalists to seasoned bureaucrats to eccentric amateurs.

Such governance issues are more than academic. It is clear that, after decades of political bludgeoning, the CBC has become a far more cautious and tamer journalistic animal than it once was. While this is much less true on the French-language side, there is an argument that English-speaking journalists not only shy away from doing the kinds of stories that cut too close to the political bone but also downplay politics entirely. Great investigative stories, exposés of political leaders and their governing styles, and discussions about ideology are increasingly hard to find in news reports. The old feedback loop, in which reporters would interact with politicians, each feeding the other stories in order to set the agenda in Question Period, has all but disappeared. In fact, there seem to be far fewer stories from Question Period. As noted in an earlier chapter, there are also far fewer political and far fewer specialist reporters than there once were.

The Trudeau government was prepared to go only partway toward emulating the process the private sector uses for board and CEO appointments. In June 2017, it named a committee of artists and filmmakers, headed by former CTV and Global broadcaster Tom Clark, to review applications from the public for board positions. Any member of the public could apply. This committee would review the applications and propose at least three names for each board vacancy, with the Heritage minister making the final choice.[30] In mid-December 2017, the government named five new directors—three to start immediately and two to begin their terms in February 2018. In an encouraging break from the past, four of the directors had either broadcasting or cultural policy backgrounds, one was Indigenous, and one came from Ottawa's high-tech and digital commerce community.

When it came to appointing the new president, however, the Trudeau government had no interest in allowing the CBC's board of directors to interview and choose Lacroix's successor. In fact, the same independent appointments committee that selected candidates for the CBC directorships had completed a review of presidential candidates and forwarded three names to the government, all of whom it rejected. The selection process was reopened to allow for applications from Canadians who were working outside the country. This wider pool included Ms. Tait, who had been working in New York. In other words, the prime minister was unwilling to relinquish the power of selection. He was unwilling to take the last step in depoliticizing the CBC, in giving up the levers of influence. Presumably, there is simply too much political value in the government having a direct line to the top management of the CBC for it to surrender that control, even if it uses it only selectively. To put it in baseball terms, Trudeau balked when it came to the crucial last batter.

At the same time that Tait was appointed president and CEO, Michael Goldbloom, a former publisher of the *Montreal Gazette* and *Toronto Star*, was named chair of the CBC board. Goldbloom is the son of veteran Liberal stalwart Dr. Victor Goldbloom. Few, however, can question his sterling credentials.

The End of the Road?

The question at the end of the rainbow is whether the combination of ignoring the CBC's blueprint for change, the implementation of Joly's Netflix policy, and the Trudeau government's reluctance to change how the president is chosen signal a final crisis for the CBC. While the CBC can sputter along for quite some time, the loss of hockey to Rogers, the evaporation of younger viewers and listeners, the advent of streaming platforms, the arrival of *peak TV*, and the need to use new platforms such as Facebook and Twitter to reach audiences mark a seismic break from the past. In the new attention economy, there is little time for the kind of endless study and introspection that we have seen in Canada, as report after report has gathered dust or been spiked for partisan reasons. There is no going back to a simpler and happier time. Most of the ground that the CBC has lost cannot be reclaimed.

The CBC, however, is not just another broadcaster in an endless ocean of other broadcasters. The vision behind public broadcasting is that it would be an essential building block in Canadian democracy and offer an alternative for a country as diverse and vulnerable to American culture and values as Canada is. The question that we all must consider is whether Canadian democracy is better off and healthier than it was 10 or 15 years ago. At the same time, are we satisfied that Canadian cultural life is thriving amid an endless cacophony of entertainment and news choices and the modern monopolies that dominate the cultural and communications landscape? If the answer is that we can do better, then the question is whether public broadcasting can be reconstructed to make a difference.

We have clearly entered the era of what Markus Prior has called "post-broadcast democracy."[31] In this new world, many of the old institutions

and assumptions need to be re-evaluated. Some of the old placehold-
ers are no longer relevant. Others will need to be reinvented. What we
propose is that, for the CBC, at least, the revolution can no longer wait.
In the next chapter, we suggest a new pathway for the broadcaster and,
indeed, for the country.

Chapter 8

Reinvent the CBC or Allow It to Die

Building a public broadcaster for the post-broadcast world begins by making some tough decisions. That starts with admitting that what has been lost is lost, regardless of why. The past is not coming back, and the future will be radically different. As we have discussed in previous chapters, the attention economy has fundamentally altered both the Canadian and the global media environments. More than that, it has left the CBC hopelessly uncompetitive in many of the fields that have been fundamental to its mandate and to its very existence.

We want to stress again that our proposals for change do not deal with Radio-Canada; they apply only to the English-language CBC. While Radio-Canada is arguably more successful than the CBC, storm clouds are also gathering around it. The forces of technological and audience changes that threaten the CBC threaten it as well.

While the digital revolution has undermined some of the roles that the CBC has long championed, it has also created opportunities—but not for a public broadcaster that tries to do and be everything to everyone. Over the past decade and longer, successive federal governments have been unwilling to say what it wants the public broadcaster to become, even as the ground has shifted dramatically around it. The problem has been that, without any direction from the government other than decisions forced through spending cuts, the CBC never wants to stop doing anything it now does. Its latest strategy document, *Your Stories, Taken to Heart*, released in May 2019, replaces 2014's five-year plan, *A Space for Us All*, with more of the expansive same. It is time to change that approach. Opportunities need to be matched and funded with money saved by stopping doing things that no longer work, where audiences have crumbled, or where it is just too costly to compete.

Simply put, to survive, a renewed and revitalized CBC must become smaller and more focused on delivering a narrower range of services to

Canadians—targeting programming areas that are most critical to its mandate and marshalling the money it has to do it. The $150 million-a-year increase for five years that the CBC has received from the Trudeau Liberals is unlikely to be increased or even replenished, particularly if there is a future change of government. There is no gold rush in the offing, no funding source not yet cashed in that is likely to save the day.

So difficult, even agonizing, decisions about what not to do, about programming limbs that may need to be amputated to save the patient, need to be made now, while there is still money to have an impact in areas where the corporation's future lies. There remains considerable potential for an organization that rebuilds some of its historic strengths in fields where opportunities have been created, not least of which by the crisis in journalism that has come with the collapse of newspaper and broadcast advertising and the hyper-competition experienced by Canada's private TV networks because of streaming and peak TV. We have to remember that the same crisis that has enveloped the CBC threatens the existence of much of Canadian media.

In effect, the CBC can flip on its head the vision, long advocated by private broadcasters, of limiting public broadcasting to the far corners of the broadcasting world—namely, to program areas that the privates have abandoned because they are too unprofitable. Private broadcasters have campaigned for decades, regularly pushing the federal government to constrain CBC television to programming that has little or no commercial interest—to what, in effect, is the advertising netherworld. Our proposal would reverse the onus. The CBC would focus on a limited number of areas, leaving private broadcasters with the responsibility to serve the country in every other arena. The CBC would, in effect, cut its losses and allow the commercial broadcasters to dominate in every other category. Any such moves would require the CBC and its employees to change their mindsets, finally abandoning the idea of competing with private broadcasters as the framework that drives CBC programming decisions. What is needed is a refocusing based on concentrating on areas where there is already an audience served by the CBC, one that has potential to grow.

The CBC can no longer do everything and be all things to all audiences. It can no longer stretch its budgets to span an entire broadcasting schedule that includes a wide expanse of programming types. It can no longer live under the burdens of a Broadcasting Act that deprives it of the ability to make strategic decisions that will allow it to survive and, indeed, prosper. The solution that we propose is that the CBC retreat from programming areas in which it can't compete, or has already lost the game, in order to throw greater resources into areas where it can make a difference. In other words, the CBC will have to do less in order to do more.

This means that the CBC must give up competing in areas such as sports and music, where specialty channels and streaming services dominate the horizon, or trying to compete against expensive mega-dramas that appear on global streaming services such as Netflix, Amazon Prime, and Apple TV+. It's a safe assumption that streaming services are growing in popularity—that they are, in fact, the new normal—and that they are becoming part of people's everyday experience. While the battle among the streaming services is difficult to predict, each of the main players has the resources to tower over anything that the Canadian media can produce. They are also able to collect and analyze data so that they can fine-tune their programming to match the interests of their subscribers.

CBC President Catherine Tait acknowledged as much in a May 2019 speech to the Chamber of Commerce of Montreal, which laid out the corporation's strategy for the next three years. In making a case for an expanded CBC, she presented a doomsday scenario as the alternative, telling her audience:

> Google, Facebook, Amazon, Apple, Spotify and Netflix already dominate the market. And they are continuing to amass more and more data on Canadians. These digital giants now know us better than we know ourselves. Through the services we use they know what we are buying. What we're reading. What we're watching and listening to. The biggest threat to our cultural sovereignty is precisely this: the appropriation of our data by foreign companies. This imbalance is keeping us from profiting from our own work. The risk is that it will smother Canadian media, arts and culture companies. It's not just a sector of our economy that is under threat: it is our identity as Canadians.[1]

A similar argument was made in the late 1980s by opponents of a free trade agreement between Canada and the United States, who warned that such a deal would critically undermine Canadian sovereignty and cultural identity as the country would be absorbed into the United States. That didn't happen in the subsequent 30 years, but in 2019, the risks Ms. Tait sees on the horizon are, in fact, already here.

The question for the Canadian TV system as a whole is, how much longer will the US producers of comedy, entertainment, and drama programming continue to carve out specific licensing deals to CTV and Global for Canadian rights to their series? It will soon make more sense, if it doesn't already, for US producers simply to sell global rights to

Netflix, Amazon, Apple, Hulu, or whatever new, over-the-top stream-ing services come along. This means that the days are numbered for special deals for Canadian TV networks that allowed them to buy shows in Hollywood at a rate far below the costs of producing a similar Canadian show. In other words, the entire edifice on which commer-cial TV in Canada is now constructed may be in jeopardy. As Richard Stursberg has noted, this would make "already unprofitable businesses even more unattractive" and allow streaming services "to crush their Canadian competitors while contributing nothing to Canadian Cul-ture."[2] As we noted earlier, these new streaming services have a large and still-growing subscriber base in Canada, so how much longer do the US program producers need the Canadians networks to distribute their programming? Perhaps not much longer at all. We believe that change is coming, perhaps sooner than many people think.

This will occur at the same time as continuing cord-cutting and the emergence of cord-nevers becomes ever more common, as households kill their cable and satellite subscriptions to specialty channels (largely owned by the same private broadcasters, CTV and Global) in favour of less expensive streaming services, which offer a wider choice and variety of programming. A discussion paper released by Ken Goldstein of Communic@tions Management Inc. in October 2018 notes that the number of cord-cutters and cord-nevers among Canadians under 30 is fast approaching 40 per cent. For those between 30 and 39, it is close to 30 per cent. These numbers are likely to mushroom to 50 per cent in the next five years.[3]

If this is indeed the case, much of the fabric that knits together Cana-dian commercial television will unravel. It may unravel in slow incre-ments, but it will unravel, nonetheless.

Although it was botched in almost every respect—from how it was communicated to specific French-language programming commitments—Heritage minister Mélanie Joly's $500 million deal with Netflix, announced in September 2017 as the centrepiece of the Trudeau government's cultural policy review, wasn't a mistake. While the fact that it seemed to dismiss the CBC in its calculations for the future was bizarre and disturbing, the policy was nevertheless a clear-eyed assessment of where we now stand. With due respect to CBC's newly launched Gem streaming service, Bell Media's Crave TV, and Radio-Canada's Tou.TV, there is no point or value in trying to create a Canadian portal and streaming service that, from the outset, won't be able to compete with the giants, either financially or in its range of pro-gramming. Add to that the crippling effects of the constraints that the CBC faces in data collection, recommender algorithms, and audience analytics, which are fundamental to the success of Netflix, Amazon Prime, and music-streaming services, among a host of others. This is the same sort of uphill battle that the CBC faced when it tried to win the rights for NHL hockey against Rogers and Bell. The results will be the same sort of losses.

The handling of Ms. Joly's announcement was clumsy, but climbing onto the global-streaming train remains one of the best options for getting Canadian stories to Canadians and the world. The question is how best to promote Canadian programming with the giant streaming services to ensure that these shows are featured prominently to Canadian subscribers and attract international attention. Government arm-twisting or regulation of the streaming services may achieve some of these goals, but the best guarantee of success is to produce excellent, timely programming that creates an audience and a sense that viewers can't afford to miss a show, no matter where they live or what language they speak. It is all about producing programs that penetrate and become a part of popular culture. The streaming success of *The Handmaid's Tale* is an example of that. To achieve that sort of standard of excellence may require spending more money on fewer Canadian programs, but it certainly means supporting Canadian programming in a different way than we have seen in the past.

The situation is very different with regard to French-language programming. As mentioned above, our analysis is limited to the English-language CBC, but it is important to point out both similarities and differences. Radio-Canada has a larger presence in Quebec, a larger silhouette, than the CBC has in English-speaking Canada, and the range of programming that Radio-Canada successfully competes in—even against Québecor's media juggernaut—is still significant. It may be possible for Radio-Canada to continue to do the range of programming that it has done to date in news, current affairs, and information as well as in entertainment, drama, and comedy. Shielded by language and Quebec's cultural nationalism, Radio-Canada has not faced the same hurricane-force winds that the CBC has. The media worlds of Canada's two languages and cultures are very different, as demonstrated by the relative success of *La Presse+*, the tablet publication that has allowed the former newspaper to stop daily printing entirely, while the *Toronto Star*'s attempt to replicate some of that with its tablet *Star Touch*, by licensing *La Presse*'s software, failed and was finally shut down. Nonetheless, the harsh economics of the new attention economy are as evident in Quebec as they are elsewhere and everywhere. Radio-Canada has arguably more breathing space than the CBC does, more time to adjust and innovate. For the CBC, however, time has run out, although the corporation continues to refuse to acknowledge that reality.

The three-year strategic plan, *Your Stories, Taken to Heart*, announced in May 2019, has at its core the argument that the public broadcaster is "A Champion of Canadian Culture." In specifics, it is similar to the CBC's 2016 expansive proposal to Mélanie Joly's cultural policy review, all of which was rejected when her Heritage department released its ill-fated plan in the fall of 2017.

What's new is the argument that the CBC should be the country's prime weapon in the fight against the FAANGs. As the document states, "Given the growing dominance of global digital companies that threaten to drown out the country's stories, as well as its news and information, we are committed to ensuring that Canadian culture thrives in the future."[4] To achieve that, the document states that the CBC will be a champion for Canadian voices and stories, will offer solutions to the growing dominance of global players, and will advocate for policy changes that ensure that digital companies make the same sort of financial contributions to the creation of Canadian content as Canadian broadcasters must now do through various levies.[5]

While the strategy paper mentions digital services, there is, surprisingly, no mention of television or radio or the specific roles that each can or is expected to play, despite the fact that neither is disappearing nearly as quickly as has been frequently predicted. Missing as well in an era of rising concern about misinformation and disinformation is anything more than a passing reference to the role that the CBC should play in news and information. The strategy paper states only what should be obvious: that the public broadcaster "will be a beacon for truth and trust against 'fake news' and algorithms that put democracy and the respect for different perspectives at risk."[6]

In fact, it seems fair to ask whether the CBC's new commitment, stated early in the document, isn't what it should already be doing, when it says, "Our promise is to put you, our audiences, first; to prioritize our role as Canada's most trusted media brand; to earn your trust and work hard to keep it every day; and to build lifelong relationships with as many Canadians as we can. We're inspired to grow our lifelong engagement with you."[7]

A New Vision Built on News and Current Affairs

Simply put, the CBC needs to give up visions of being the primary outlet for telling Canadian stories to Canadians on television in drama, entertainment, and comedy. It played an important and valuable role in the past, one that helped build a Canadian television-production industry, in part by deciding to eliminate its own in-house production capabilities. Today the public broadcaster is solely a commissioner, buyer, and distributor of privately produced Canadian programming. However loyal the audience for shows such as *Murdoch Mysteries, Kim's Convenience*, or *Workin' Moms* may be, we seem destined to soon reach the point where no show can be truly successful or popular unless it is also carried by Netflix, Amazon Prime, Hulu, or one of the other streaming goliaths. With increasingly older audiences, and without the budgets or promotional firepower needed to compete against the avalanche of original shows that are produced every year in the new attention economy, most of which end up on streaming services, Canadian shows will have difficulty entering the mainstream of popular culture.

But there is an alternative, another way forward. Perhaps the best analogy for what we believe should happen lies in sports and, specifically, Canada's preparations for the 2010 Vancouver Olympics. It was an opportunity to showcase Canada and its athletes to the world and also avoid a repeat of the embarrassing inability of Canadian athletes to win a single gold medal at either the Montreal summer games in 1976 or Calgary's winter games in 1988.

To prevent a hat trick of failure in Vancouver, the federal government created a new organization and gave it a distinctly un-Canadian name: Own the Podium. The goal was simple yet, for Canada, wildly audacious—to become the best in the world and win the most gold medals of any nation competing in Vancouver. For some, the approach seemed a shocking abandonment of the meek-and-mild Canadian sports persona: athletes always trying hard but in the spirit of good sportsmanship ready and willing to concede publicly that someone else was better and being more than satisfied to perform at their personal best while losing.

Own the Podium marked a crucial shift in psychology. It challenged some entrenched assumptions. Why can't that winning performer be Canadian? What is stopping this from happening, and why should athletes be satisfied with anything less than the top step on the podium? Striving for excellence became the message.

The hoped-for swagger of Own the Podium made many Canadians uncomfortable. But the desire to win resonated among young people and seemed natural, as the growing crowds in the streets of Vancouver demonstrated, as the games progressed and Canada was suddenly battling for the top spot in gold-medal standings. The millennial generation has grown up comfortable with and proud of their Canadian identity, confident that Canada is a country that the world envies. They see nothing wrong with saying that they aspire to be the best in the world. That is equally true of young French- and English-speaking Canadians. The era of living in the shadow of American corporate, economic, and even cultural domination, as a cultural suburb of the United States, seemed to have faded, at least when it came to the Winter Olympics. (As an aside, the vision laid out by Ms. Tait, the CBC president, which viewed the FAANGs as threats not just to a sector of the Canadian economy but also to Canadian identity, seemed to come straight from that old, defeatist perspective.)

By the Olympics in PyeongChang in 2018, the transformation had reached the point where Canadian athletes and the Canadian public expected to win. They were deeply disappointed when Canadian athletes failed, as happened in curling and hockey, but there were many achievements worth celebrating. Canada finished third in medals overall, a staggering accomplishment.

The success of Own the Podium came from a significant increase in direct government funding for sport, initially for winter sports, but then expanded for summer games as well. Government wanted sport success, so it funded it directly. Own the Podium and individual sports also obtained private-sector funding, but that complemented the money that came from the federal government for athletes, sports organizations, competitions, equipment, research and science, and training and competition facilities constructed for past Olympics in Calgary, Vancouver, and more recently, the Pan American and Parapan American Games in southern Ontario.

We use the example of Own the Podium to suggest that great things can be achieved if you aim high enough and are brave enough and that picking your spots, being strategic in what you strive for, can be decisive. Just as Own the Podium specialized mostly in winter sports, recognizing that this is where winning was possible, we believe that a stripped-down CBC that focused on being excellent in just a few areas would ensure its survival and be a valuable resource for the country.

If promoting Canadian media and Canadian news is a national priority, then government will have to play a major role, especially given the fact that the stakes have increased considerably with the emergence of global media platforms with almost unlimited audiences and resources. To put it differently, in the vast new attention economy, we believe that it will take a national commitment to ensure that Canadians see their own media and cultural reflection and receive the quality news that is the spinal cord of any democracy. Instead of watching the CBC slide into irrelevancy, we think that it is now time for the government to revive and restore public broadcasting. This time, the CBC would be far different from the one that previously existed.

It is not clear to us that the days of formidable public broadcasters are over. In Europe and through much of Asia, public broadcasters continue to play a decisive role in their societies. They also attract large audiences, despite the splintering brought about by satellite services, with their myriad of specialty channels and the Internet. The BBC is perhaps the best model of that, as the public face of the United Kingdom, its society, its values, and its national perspective on the world around it. Public broadcasters in Germany and in the Scandinavian countries continue to stamp their signatures on the political and cultural lives of those societies. It's also the case, as mentioned previously, that, in many countries, public broadcasters have led the way in adapting to digital change. What we propose is that the CBC assume that role in Canada by choosing battlefields on which it can make an important impact. The question is, what does the country need from a public broadcaster in order for its political and cultural life to succeed amid the hurricane-force winds of the new attention economy?

What we believe is required is a bold restructuring of the mandate, role, and place of the CBC in Canada's media universe. As we stated at the beginning of this chapter, this involves conceding that what is lost, is lost. The CBC cannot recapture its past glories in sports, music, children's TV, or, sadly, in comedy and drama. This means jettisoning entertainment programming and allowing it to move to streaming services. Entertainment would fall within the ambit of private producers and broadcasters and be supported directly by government, not through indirect measures such as the Canada Media Fund. The CBC argues that it should play a brokerage role, as Ms. Tait told her Montreal audience: "Becoming a spearhead for Canadian content in international markets will become a key objective for our next strategic plan."[8] Yet the CBC has, at best, very limited experience brokering Canadian programming to international streaming services, and that role was rejected by the Heritage department in 2017. One can imagine instead an enhanced role for the National Film Board as the broker of Canadian programming, directly funded by government, to the streaming services as well as overseeing co-production agreements with global partners. Some shows are likely to do well—although the key is likely to be deals with streaming behemoths such as Netflix and Disney. The CBC will no longer be in that game.

While the CBC may wish to cling to the fantasy that it is still a major player in sports because it airs two NHL games on Saturday nights featuring Canadian teams, the reality is that Rogers's deal with the NHL, and its side deal with Québecor for French-language rights, effectively knocked the CBC off the ice. The charade can be maintained for only so long. However laudable its decision to carry emerging leagues such as the Canadian Premier Soccer League, the IAAF Diamond League track and field, and FIVB Nations League volleyball, these events can be offloaded to sports specialty channels, presuming that they even want them.

News and current affairs, however, are another matter. News remains the fulcrum and lifeline of democracy. Reliable and objective information, combined with well-reasoned and cogent analysis, is essential if citizens are to have an understanding of the world in which they live. The CBC cannot retreat from being a main news provider. In fact, this role needs to be reinforced and become the tip of the spear for a new CBC. The backdrop is, of course, the crisis that is enveloping the Canadian news media generally. As the Canadian newspaper industry plunges into a death spiral from which it may never pull out, and local TV news shows mounted by the private networks continue to lose advertising and audiences, saving the news has become a national priority.

There can be no great countries without great journalism, the kind of journalism that exposes issues, stirs thinking and debate, and produces credible facts and information. Too many news organizations promote opinions and ignore expertise; have few, if any, specialist reporters; crave sensational stories at the expense of accountability reporting; and do little, if any, digging or investigative work. They do not serve citizens well. Although snowflake news, which disappears almost as soon as it lands because it has no lasting value or impact, and news based on the latest antics of Justin Bieber or Kanye West may pay dividends in the short run, we have seen citizens become increasingly distrustful and skeptical of much of the news that we receive. The new CBC would provide citizens with news and current affairs programming and online content that goes beyond the fluff and provides citizens with "must-have" news. Having shed its other responsibilities, the CBC will now have the resources to be truly excellent in the one area most crucial to the health and future of democracy.

This does not mean that the CBC can achieve success easily or that its path will not be filled with obstacles and dangers. A main difficulty is that many, and perhaps most, people now receive their news through smart phones and, as a result, have become used to so-called news snacks—short headlines, prompts, posts, video hits, etc.—and have little patience for anything more—except on a handful of key issues that are important to them. It is also the case that filter bubbles and feedback loops endlessly play back people's preferences and prejudices. Where

TV news and newspapers once shaped the views of their customers, today people's identities determine the news outlets that they choose. Matthew Hindman refers to this as "lock-in"—and for many people, once they are in lock-in mode, the psychological and search costs of changing channels or news sites are too taxing.[9]

The challenge for the CBC is to become part of the me-media echo chambers of a large number of Canadians—a task that demands extraordinary effort and resources. Writing and producing for different platforms is extremely demanding because the affordances and audience structures of Facebook, Twitter, Instagram, WhatsApp, and Snapchat, etc. are all different. Some are more populist, some more visual, some more open to outside stories, and so on. Failure on this front can be life-threatening for any news organization because each of these platforms has the capacity to downplay or ignore its stories. How news organizations play the game of platform politics may determine how and whether they survive.

The CBC also has to take a hard look at where its current obsession with Chartbeat is leading it. While Chartbeat can be a valuable tool, public broadcasting cannot become another version of *BuzzFeed*—with clickbait headlines and triggers that guarantee that stories go viral. Designer stories are not what public broadcasting should be about.

What we are suggesting is a news organization that is the essential touchstone for people who need high-quality information about Canada's place in the world, Canadian business, advances in health care and science, and Canadian culture and will simply not find what they need conveniently in other places. We are also suggesting a heavy dose of accountability news—reporting about how institutions function, covering the front lines of political and civic life, and heightening the quality of public debate. The CBC has to flood the zone with news that people "must have"—and "must turn to" for professional knowledge, to be good parents, to be in business, to be an engaged member and participant in the life of their communities, and to know about the world.

The CBC's new mandate does not let private TV stations and newspapers off the hook. In fact, their responsibilities will increase. As the CBC withdraws from entire programming areas, private broadcasters will have those fields to themselves. We suspect that they will also increase their commitment to news because news shows, particularly at the local level, are signatures of community branding and involvement. While commercial media outlets have cut news staffs and retreated from the front lines of investigative news, they may have to invest in their news budgets to keep the attention of their readers and viewers. At the same time, newspapers will likely continue to struggle, and it is likely that a number of the old lions of Canadian journalism will fade away or be weakened to the point where there is little left—and a renewed CBC may provide the country with an insurance policy as newspapers change their form. Some, of course, will succeed.

All this requires that the CBC undergo a dramatic change of direction—that it wear a new face. Although it received more money for its digital transformation from the Trudeau government, little of that money is going to news and current affairs. The push to go digital has come out of the news department's own budgets, and those budgets are now being cut, even in the face of growing concerns about the deteriorating quality of news and the threats that this poses to Canadian democracy. While the CBC faces major challenges, especially with regard to the faltering performance of *The National* as well as its decisions over the years to weaken local TV news, the CBC still maintains a solid news backbone in radio and is a leader in online news. For quite some time, CBC News has attracted the largest digital audiences of any Canadian news organization—although, as pointed out in chapter 4, its user base is still small compared to any of the major social media sites. Moreover, CBC News continues to explore opportunities on different platforms such as Snapchat and Amazon's virtual assistant, Alexa; these may or may not have long-term appeal, but they indicate that the CBC is actively investigating new partnerships so that it can take advantage of the next digital wave.

Nor can it continue to abandon local communities. It has to reverse the disastrous decisions taken in the early 1990s and after to variously shut down, weaken, gut, move, revive, and then try to restore local TV news shows. Moreover, its drive-time radio shows are still important meeting places in many Canadian communities, and they should not be subject to any more bloodletting.

It is important to note that CBC News has also had successes with audience engagement through such features as Vote Compass, which, in each federal and provincial election, surveys online visitors about their political views and, based on their responses, calculates which party in an election advocates policies closest to what they believe. For instance, in the 2018 Ontario provincial election, Vote Compass was used more than 365,000 times on the CBC News website.[10] Unlike many news organizations that have eliminated online comments in the face of non-stop barrages from trolls and extremists, CBC maintains mediated comment sections on its news pages as a way of trying to engage with its audiences. In total, local and network news and information represents 87 per cent of the CBC's digital audience—a strong signal about where the digital future for the public broadcaster actually lies.

Strangely, the focus of the CBC's spending on its digital transformation is mostly on entertainment and drama and not on news, information, or current affairs. This is a fundamental and perhaps life-threatening mistake. Failing to build on obvious strengths and past successes, while propping up areas where it is unlikely to be able to compete successfully, is mystifying.

The Revolution Won't Have Advertising

Another big step is to end all advertising. Leave that declining source of revenue to private media organizations to fund their operations. That one change will dramatically affect everything the public broadcaster does, freeing it from its current dominant mentality, in which it views itself as being in constant competition with the private sector, and for that reason consistently (until very recently)—and, we believe, foolishly—rejecting any suggestion that it should co-operate with other media organizations except in very limited ways. A by-product of a decision to eliminate all advertising would make the CBC more attractive to younger audiences, who appreciate and are now used to the lack of ads on streaming services such as Netflix. Eliminating advertising would also change programming decisions as programmers would no longer have to take into account how interruptions for advertisements distort the stories they want to tell.

How different might the relaunch of *The National* have been in November 2017 had the broadcaster announced that there would be no advertising on the hour-long newscast? That would have attracted a curious audience in itself and would also have given the program's producers tremendous flexibility in how and what they presented when not constrained by the regular advertising interruptions that disrupt any sort of programming continuity and offer viewers invitations to leave. This last point is crucial. In the new attention economy, another channel, program, platform, video, or post is just a tap away. Halt the action for just a minute, and your audience may be gone.

As discussed in chapter 4, the art form today is stickiness. The key to success for Facebook and Netflix is that they have a sticky quality: they keep users glued for as long as possible. The CBC seems to be going in the opposite direction. Instead of eliminating advertising on its flagship newscast, the CBC has increased the number of ads on *The National*.

The obvious point of comparison is CBC Radio, where the decision to end advertising decades ago has created a distinctive and popular service, nothing like its private-sector competitors. This explains to some degree, at least, why it has the largest single share of listening audience in most metropolitan communities across the country. That CBC TV could replicate radio is perhaps too much to expect. Putting an end to advertising, however, would change almost everything the CBC does.

In addition, it is not as if the broadcaster would be giving up a gold mine of future riches by doing it. They have very little to lose—so why not take the plunge?

As mentioned above, another point of comparison is with Netflix and the other streaming services. One of the ways in which Netflix has changed the culture is that it has made Canadians used to viewing without advertising. The absence of advertising allows producers to write longer scripts without the artificial peaks and valleys created by ads. It also allows them to cross boundaries in terms of taste and violence that might have chased conservative advertisers away. One of the reasons for the triumph of serial storytelling is precisely the absence of advertising.

An end to advertising will also allow CBC employees to think differently about news, information, and current affairs. Instead of just reporting the news, the public broadcaster should spend more time and attention on finding and creating the news. This requires a fundamental reconceptualization of what news is and should be and how it is gathered and presented. For the CBC and its employees, it means returning journalists to the centre of what the public broadcaster does. The focus should become building and showcasing their expertise, experience, and knowledge, instead of having "journalists" spend their days sitting at desks like bureaucrats, reading and regurgitating social media content in the pursuit of clicks. Without the need to impress advertisers, there would be less need to use numbers of clicks as the standard of success. The obsession with clicks would end. Putting journalists back on the front lines might also have the effect of attracting a new generation of journalists and, in the process, reinvigorating the entire profession.

Constructing a New Identity

It is also time to rethink the CBC's online presence. Just as getting out of advertising allows those producing television programming to start thinking and making decisions without worrying about how this will affect advertising and what their private-sector competition is doing, so too can online be rethought. Use the CBC name and presence online to help build, not compete with, local, private-sector media. Instead of entrepreneurs seeing the CBC as a competitor or a deep-pocketed rival that can drive them out of business, the CBC should be their partner.

Share stories, give local media provincial and national prominence by featuring their work on CBC online sites, and encourage the public to subscribe to private-sector media outlets to keep them going. Turn the CBC local news and information sites into innovators, working with small news organizations and journalism and communication students to bring their stories to life. New online media outlets such as *iPolitics*, *The Tyee*, *The Conversation*, *National Observer*, and *J-Source* already swap stories freely, giving each other credit when they run on someone else's site.

We also mentioned the new focus on crowd-based journalism. Reporters let readers see what they have found and invite them to contribute information they have that might advance the story. Again, the *Washington Post* used "the crowd" to discover new facts about Donald Trump's charitable donations and won a Pulitzer Prize as a result.

While constructing such a new online highway is audacious, even revolutionary, the country may have little choice. All the major FAANG platforms exist beyond Canada's borders and beyond Canadian control. They are making the key decisions about the newsfeeds that Canadians receive. If, as some suggest, the entire Canadian news infrastructure is in jeopardy, then collaboration may be the last great hope. What better goal for public broadcasting than to take the lead in championing the needs of the entire Canadian media system? The days of fighting to the death—and watching everyone die—is no longer sustainable for a country like Canada.

Finally, the CBC's online news sites should encourage news literacy as a core part of their mandate. Such a move could include promoting a greater understanding of the standards that news organizations use to decide what gets on the air or online. Teaching people how to distinguish between fact and fiction, between comment and analysis, and between errors and deliberate misrepresentation or falsehood is important—not just for elementary or high school students but also for all audiences.

We are now in an era when the curating role that news organizations used to play—acting as gatekeepers that did that sorting for audiences—no longer exists. Social media, with its "interesting if true" standards for accuracy, has replaced mainstream media for many, if not most, Canadians. Everyone has the ability to create their own highly personalized media echo chamber and be their own news curator, sorting through information and recommendations on social media, on online sites, in stories forwarded from friends, and in material posted by groups anxious to stir up social unrest. Everyone needs to know how to navigate through that world successfully—a job the public broadcaster can embrace in the spirit of community education.

While the CBC needs to be radically reinvented on TV and online, this is not the case with radio. It should continue to do what it has done successfully for decades, only with the benefit of more money and, as we have mentioned above, an end to the cutbacks that have handicapped it. Radio continues to play a vital role in communities across the country, and it should be a mainstay of the public broadcaster's local presence. Allow radio to return to independent storytelling, no longer forced to use audio tracks of television stories in its newscasts. Also, eliminate the overlap and duplication, often to the point of using the same stories on both the main radio newscast, *The World at Six*, and television's *The National*. Recognize that podcasts are, at their heart, radio programming—interviews and storytelling in a personalized and more intimate format that responds to the audience's desire to listen when it wants to or has the time to do so rather than when the broadcaster decides to air them.

Where does the overhaul that we advocate leave CBC News Network? Do cable and satellite news channels have a long-term future? Almost certainly not. Audiences are already deserting them, just as they are leaving mainstream television, and all-news audiences were never that large in the first place. Still, since cable has situated itself as the gateway to streaming services, we should be careful about writing its obituary just yet. While cord-nevers and cord-cutters are a growing reality, pulling the plug on CBC News Network may be premature. Having a second window for election nights, Supreme Court decisions, fast-breaking news stories, and international crises may be valuable, especially as a new CBC takes hold. Most important, perhaps, is that the subscriber income that the CBC receives from its News Network is now the single largest source of revenue for the corporation outside of its parliamentary appropriation, a perch once occupied by income from *Hockey Night in Canada*.

One trap that is all too easy for news channels to fall into is the artificial shock and awe of so-called big-event news. All-news channels going back to the first Gulf War in 1990 have discovered that big events bring people flocking to them to watch, such as the first US missile attacks on Baghdad, famously covered by CNN, with reporter Peter Arnett live in Baghdad. Once the big event ends, however, the audience quickly melts away and shrinks to its pre–big event size. Who knows, as well, how long it might be before the next big event comes along. That unpredictability has led news channels to try to create big events, turning the mundane into something dramatic, with non-stop banners on the screen screaming that a new development is coming up right after the next commercial, while endlessly chewing over the most trivial aspects of a story or event in the hope that something else will happen. It is not hard to see that the pursuit of big events can easily become farcical.

Shifting the News Lens

We suggest that the new CBC must focus on a specific set of issues important to Canadians, reported by Canadians, in ways that relate to Canada and the life experiences of Canadians.

So what should that focus be? We would describe it as ensuring that there are Canadian eyes observing, reporting, and analyzing the world around them and explaining the impact that those worlds will have on individual and collective lives, communities, and experiences across the country. There are several areas that we believe a public broadcaster should concentrate its attention on in the post-broadcast world, whether it be on radio, television, or online.

First is interpreting the world for Canadians. The private-sector media has almost completely withdrawn from covering the world through the eyes of Canadian journalists, substituting content provided by foreign news organizations and their journalists, which may or may not relate to Canadian experiences. Yet interpreting global developments, opportunities, and risks for Canadians has never been more important. Thanks to changing technology—as the CBC has itself demonstrated, with its pop-up news bureaus in Istanbul (now closed) and Moscow (now once again a permanent CBC bureau)—it has never been easier to report from anywhere, even the most isolated places on the planet. There should be a significant expansion in the number of CBC journalists based around the world. It is not good enough to have a handful of journalists clustered in one or two capitals, from which they occasionally go out to cover natural disasters or major political, social, or economic crises but are unable to report on daily life and the challenges that societies face in nations where so many Canadians once called home and still have relatives. Spreading Canadian journalists around the globe can help build real links, which now barely exist between the public broadcaster and Canada's diversity of ethnic communities.

Most crucially, this also means a greater presence in the United States beyond Washington, New York, and Los Angeles: in the cross-border regions and communities that most directly interact with and affect Canada and Canadians. Events there since 2016 have demonstrated the degree to which Canadians actually know very little about our closest neighbour. The public broadcaster should lead in closing that gap in our collective knowledge and our understanding not just of politics in Washington, entertainment in Los Angeles, and the United Nations in New York.

To complement a significantly expanded network of Canadian foreign correspondents, local radio and online news and national television reporters located in communities across Canada should sharply narrow the range of domestic issues they cover. The CBC should tell its audiences that if they are looking for entertainment, fashion, lifestyle, sports, crime, traffic accidents, weather, and the latest crazes, lunacies, and viral outbursts on social media, they should look elsewhere. After all, it *is* everywhere. Let the private sector cover it all for as long as it can find the advertising to support it.

In its news and current affairs coverage and programming on radio, on television, and online, the CBC should concentrate on a narrower range of subjects and themes. We think the public broadcaster should focus its news, current affairs, and information around five themes:

- **Urban life:** Everything from transportation, urban growth, and environment to social policy, immigration and diversity, housing, and community development across Canada and abroad.
- **Business and the economy:** This affects all Canadians, and they need journalism that explores and explains the issues, challenges, successes, and failures using language and approaches that allow Canadians to understand how these issues affect them, their families, and communities, whether they are rich or poor, urban or rural. This theme also encompasses jobs, technology and the future of work.
- **Public policy at the federal, provincial, and municipal levels:** Less about the politicians and more about the impact on Canadians in all demographic groups across the country of what politicians propose and do. A focus here should be more on exploring solutions to problems, if possible, before a problem reaches a crisis stage. Linked to that is a greater willingness to compare and contrast how other countries, provinces, and communities respond to specific problems, showing the different approaches that may exist that achieve success.
- **Health and science:** Avoiding the breathless study-of-the-week and cure-of-the-week reporting to concentrate on research about patient issues, delivery of health care services, health promotion, and all forms of scientific exploration and discovery.

- **Canadians who are making a difference:** Telling the stories of
  individual Canadians—their ambitions, goals, challenges, successes,
  and failures across all sorts of activities and all walks of life, both in
  Canada and around the world. While Canadians may be familiar
  with the names of people who make headlines, write books, act in
  movies, are athletes, artists, or thinkers, or are emerging political
  stars, they often know little about their life stories, motivations, and
  beliefs. The CBC can help tell these stories.

This focus should also be the foundation for local news coverage.
These five themes would be the overlay around which local news is
constructed. While local news from the private sector can keep citizens informed about police chases, traffic accidents, crime and the
courts, sports, and weather, the CBC's scope will be wider. It will stress
accountability news by covering city halls, school boards, and the func-
tioning of the health care system and dive deeply into cultural events.

The goal, again, is to produce "must-see news" that will be an important part of the national conversation. Presumably, it will be news that
government and community leaders, business people, educators, and
citizens will require to make decisions about our lives. In the hyper-
competitive attention marketplace, such news will have cleared out a
distinctive space: news about Canada. The result would bring a dramatic change to the Canadian media landscape and a clear differentia-
tion between the public and private sectors, effectively extending what
has existed in radio for several decades to both television and online
news. It would both narrow and clarify the mandate and focus of the
CBC and carve out a distinct place in the media environment for public
broadcasting.

The new vision would mean a sharp psychological break from the
past, just as Own the Podium challenged Canada's sports system to
change by becoming bolder in its goals and its expectations, by raising
the bar. As we have said, the past is not coming back, nor will whatever
success was achieved in that past translate well to the years ahead. We
believe that this new, more focused direction for public broadcasting
in Canada will help ensure that it has a prominent place and role, both
despite and because of the digital revolution. Although the delivery
systems for news and information have changed and will continue
to change, the importance of public broadcasting in Canada remains
undiminished, despite its vocal critics. The challenge is how to sustain
public broadcasting when everything all around it is changing. We
believe that the answer is to focus on the traditional strengths of the
CBC and public broadcasting more generally. We are confident that

there remains a strong and growing audience for a renewal of public broadcasting in a world drowning in information but critically short of facts and context.

A key part of the equation, however, is a new scheme for financing and governing the CBC. For almost the entire history of the CBC, political leaders have tried to use it as a political football. Threatening, cajoling, criticizing, monitoring, boycotting, and attempting to control the public broadcaster have marred and interrupted much of its progress. Governments have used at least three mechanisms of control: the carrot and stick of budget allocations, the CBC's board of directors as a kind of Senate filled with party loyalists and donors, and the power to name the corporation's president and CEO. However tempting it might be for politicians to want to control the purse strings and yank them from time to time, the CBC will need the protective cover of long-term funding if it has any hope of embarking on a new vision.

While it is clear to us that the licence fees used to finance public broadcasters in most other countries have led to greater success and have managed to shield them from the long arm of political interference to a greater degree than has been the case in Canada, introducing such fees in Canada is likely to create a sharp backlash. A five-year appropriation by Parliament would allow long-range planning and provide some measure of political protection. We also believe that the board should be sheltered as much as possible from the helter-skelter of political battle. Ideally, appointments to the board might come from a neutral body made up of citizens with suitable credentials, and the president and CEO would be chosen by the board members based on their experience and, indeed, their political neutrality.

Without radical change, we believe that the CBC's days are numbered and that the entire Canadian media and journalistic infrastructure might be in jeopardy for many of the same reasons. The CBC will survive only if we have the courage to make bold decisions.

Notes

1. Introduction

1 Brooks DeCillia and Patrick McCurdy, "The Sound of Silence: The Absence of Public Service Values in Canadian Media Discourse about the CBC," *Canadian Journal of Communication* 41, no. 4 (2016).
2 Brooks DeCillia and Patrick McCurdy, "Viewing the CBC as a Public Good," *Policy Options* (Montreal: Institute for Research on Public Policy, 24 November 2016).
3 Christina Holtz-Bacha, "The Role of Public Service Media in Nation-Building," in *Public Service Media in Europe: A Comparative Approach*, ed. Karen Arriaza Ibarra, Eva Nowak, and Raymond Kuhn (London: Routledge, 2015), 27–40.
4 Trine Syvertsen et al., *The Media Welfare State: Nordic Media in the Digital Era* (Ann Arbor: University of Michigan Press, 2014), 19.
5 John Meisel, "Escaping Extinction: Cultural Defence of an Undefended Border," in *Southern Exposure: Canadian Perspectives on the United States*, ed. David Flaherty and William McKercher (Toronto: McGraw Hill Ryerson, 1982), 152.
6 John Ralston Saul, quoted in Chris Cobb, "Saul Enters CBC Debate," *National Post*, 30 January 2001, A10.
7 Benedict Anderson, *Imagined Communities: Reflections on the Origins and Spread of Nationalism* (London: Verso, 1983).
8 Josianne Millette, Mélanie Millette, and Serge Proulx, *Attachement des communautés culturelles aux médias: Le cas des communautés haïtienne, italienne et maghrébine de la région de Montréal*, Cahiers-médias no. 19 (Sainte-Foy, QC: Centre d'études sur les médias, Université Laval, 2010).
9 Herbert Simon, quoted in Matthew Hindman, *The Internet Trap: How the Digital Economy Builds Monopolies and Undermines Democracy* (Princeton, NJ: Princeton University Press, 2018), 4.

10 James Webster, *The Marketplace of Attention: How Audiences Take Shape in a Digital Age* (Cambridge, MA: MIT Press, 2014); Tim Wu, *The Attention Merchants* (New York: Knopf, 2016).

11 See Webster, 49–53.

12 Herbert Simon, quoted in Hindman, *The Internet Trap*.

13 David Taras, *Digital Mosaic: Media, Power, and Identity in Canada* (Toronto: University of Toronto Press, 2015).

14 Henry Jenkins, Sam Ford, and Joshua Green, *Spreadable Media: Creating Value and Meaning in a Networked Culture* (New York: New York University Press, 2013). The phrase comes from the title of a white paper, "If It Doesn't Spread, It's Dead," by Henry Jenkins et al. (Convergence Culture Consortium / MIT, 2008). It also appears on the back cover of *Spreadable Media*.

15 Paul Rutherford, *When Television Was Young: Primetime Canada, 1952–1967* (Toronto: University of Toronto Press, 1990), 134–45.

16 *CBC/Radio-Canada Annual Report 2017–2018* (Ottawa: CBC-Radio-Canada, 2018).

17 Barry Kiefl, "On Treacherous Ground," *Literary Review of Canada* 22, no. 5 (June 2014): 21.

18 Canadian Radio-television and Telecommunications Commission (CRTC), *Communications Monitoring Report 2018* (Ottawa: CRTC, 2019), 209.

19 Frederick Bastien, *Breaking News: Politics, Journalism and Infotainment on Québec Television* (Vancouver: UBC Press, 2018), 24.

20 *CBC/Radio-Canada Annual Report 2017–2018*.

21 CRTC, *Communications Monitoring Report 2018*, 228.

22 Ibid.

23 Linda Stone, "Just Breathe: Building the Case for Email Apnea," *Huffington Post*, 2 August 2008.

24 Webster, *Marketplace of Attention*, 16.

25 Adam Alter, *Irresistible* (New York: Penguin, 2017), 15.

26 Patricia Cormack and James F. Cosgrave, *Desiring Canada: CBC Contests, Hockey Violence, and Other Stately Pleasures* (Toronto: University of Toronto Press, 2012); Beverly J. Rasporich, *Made-in-Canada Humour: Literary, Folk and Popular Culture* (Amsterdam: John Benjamins Publishing, 2015).

27 Communic@tions Management Inc. "Betting on the Games: Canadian Television's Annual Spending on Sports Programming Now Exceeds $1 Billion a Year," discussion paper, 14 March 2019.

28 Ibid.

29 CRTC, *Communications Monitoring Report 2018*, 208.

30 Ibid.

31 Ibid.

32 Canada, Parliament, Senate, Standing Senate Committee on Transport and Communications, *Time for Change: The CBC/Radio-Canada in the Twenty-First*

Century, Report of the Standing Senate Committee on Transport and Communications, July 2015, 50.

33 Richard Stursberg, *The Tower of Babble: Sins, Secrets and Successes inside the CBC* (Vancouver: Douglas & McIntyre, 2012), 117.

34 CBC/Radio-Canada, "Thank You Canada! CBC/Radio-Canada Doubles Its Digital Reach Two and a Half Years Early," news release, 11 December 2017, https://www.newswire.ca/news-releases/thank-you-canada-cbcradio-canada-doubles-its-digital-reach-two-and-a-half-years-early-663415263.html.

35 Barry Kiefl, "CBC's Future Should Look Like Its Radio Past," *Globe and Mail*, 6 January 2014.

36 Wade Rowland, *Canada Lives Here: The Case for Public Broadcasting* (Westmount, QC: Linda Leith Publishing, 2015), 169.

37 Phil Ramsey, "Public Service Media Funding in Ireland Faces Continuing Challenges," *Media Policy Project* (blog), London School of Economics, 26 September 2017; RTÉ, "RTÉ Broadcasts 15 of Top TV Programmes of 2015," news release, 7 October 2017.

38 Syvertsen et al., *Media Welfare State*, 12.

39 Communic@tions Management Inc., "Canada's Digital Divides," discussion paper, 20 August 2015.

40 Communi@tions Management Inc. "Betting on the Games."

41 Public Policy Forum, *The Shattered Mirror: News, Democracy and Trust in the Digital Age* (Ottawa: Public Policy Forum, January 2017), 28, https://shatteredmirror.ca/.

42 Elizabeth Grieco, "Newsroom Employment Dropped Nearly a Quarter in Less than 10 Years, with Greatest Decline at Newspapers," Pew Research Center, 30 July 2018.

43 CRTC, *Communications Monitoring Report 2018*, 240.

44 Public Policy Forum, *Shattered Mirror*, 27.

45 See, e.g., Christie Blatchford, "Mainstream Media Is Starving," *National Post*, 30 June 2018, A4.

46 Knut Erik Holm, "The Debate about Public Broadcasting in the Modern Digital World: A Case Study of NRK," research paper (Oxford: Reuters Institute for the Study of Journalism, University of Oxford, 2014).

47 Virginia Heffernan, *Magic and Loss: The Internet as Art* (New York: Simon & Schuster, 2016), 15.

2. Lost Horizons

1 Amanda Lotz, *We Now Disrupt This Broadcast: How Cable Transformed Television and the Internet Revolutionized It All* (Cambridge: MA: MIT Press, 2018).

2 David Hendy, *Public Service Broadcasting* (Basingstoke: Palgrave Macmillan, 2013).

3 Knowlton Nash, *The Microphone Wars: A History of Triumph and Betrayal at the CBC* (Toronto: McClelland & Stewart, 1994), 49. For an excellent history of Canadian broadcasting policy, see Marc Raboy, *Missed Opportunities: The Story of Canada's Broadcasting Policy* (Montreal and Kingston: McGill-Queen's University Press, 1990).

4 Nash, *The Microphone Wars*, 49.

5 Ibid., 85.

6 Raboy, *Missed Opportunities*, 247.

7 Nash, *Microphone Wars*, 51.

8 Hendy, *Public Service Broadcasting*, 7–26.

9 See Raboy, *Missed Opportunities*, 22–30.

10 Nordicity, "Analysis of Government Support for Public Broadcasting," prepared for CBC/Radio-Canada, 11 April 2016, 13.

11 Patrick Watson, *This Hour Has Seven Decades* (Toronto: McArthur and Company, 2004), 243.

12 Paul Rutherford, *When Television Was Young: Primetime Canada, 1952–1967* (Toronto: University of Toronto Press, 1990), 491–92.

13 David Taras, "The Mass Media and Political Crisis: Reporting Canada's Constitutional Struggles," *Canadian Journal of Communication* 18, no. 2 (Spring 1993): 131–48.

14 Nash, *Microphone Wars*, 151–52.

15 Mary Vipond, *The Mass Media in Canada* (Toronto: James Lorimer, 2000), 38.

16 For an impressive history of the CBC during the war years, see Connor Sweazey, "Broadcasting Canada's War: How the Canadian Broadcasting Corporation Reported the Second World War" (master's thesis, University of Calgary, 2017).

17 Florian Sauvageau, "Millennium Blues: The 1997 Southam Lecture," *Canadian Journal of Communication* 23, no. 2 (1998): n.p.

18 See Michael Grossman and Martha Kumar, *Portraying the President: The White House and the News Media* (Baltimore: Johns Hopkins University Press, 1981); Thomas Patterson, *Out of Order* (New York: Knopf, 1993).

19 F. Christopher Arterton, "The Media Politics of Presidential Campaigns," in *The Race for the Presidency: The Media and the Nominating Process*, ed. James David Barber (Englewood Cliffs, NJ: Prentice Hall, 1978), 3–25.

20 See David Taras, *Power and Betrayal in the Canadian Media* (Peterborough, ON: Broadview Press, 2001), 131–33.

21 Ibid., chap. 5.

22 Ibid., 142.

23 Ibid., 158.

24 Nicholas Russell, *Morals and the Media: Ethics in Canadian Journalism* (Vancouver: UBC Press, 2006), 88.
25 Taras, "Mass Media and Political Crisis."
26 David Taras, "The Struggle over *The Valour and the Horror*: Media Power and the Portrayal of War," *Canadian Journal of Political Science* 28, no. 4 (December 1995): 725–48.
27 Ibid., 137.
28 Watson, *This Hour Has Seven Decades*, 484–85.
29 Wade Rowland, *Saving the CBC: Balancing Profit and Public Service* (Westmount, QC: Linda Leith Publishing, 2013), 92.
30 Lotz, *We Now Disrupt This Broadcast*.
31 Gregory Taylor, "Canadian Journalism Is in Crisis: CBC News Can Save It," *Huffington Post*, 28 June 2017.

3. The Politics of Resentment and Neglect

1 Wade Rowland, *Canada Lives Here: The Case for Public Broadcasting* (Westmount, QC: Linda Leith Publishing, 2015), 198.
2 See chap. 3 in this book.
3 Bank of Canada, inflation calculator, https://www.bankofcanada.ca/rates/related/inflation-calculator/.
4 Nordicity, "Analysis of Government Support for Public Broadcasting," prepared for CBC/Radio-Canada, 11 April 2016.
5 David Taras, *Digital Mosaic: Media, Power, and Identity in Canada* (Toronto: University of Toronto Press, 2015), chap. 2.
6 Antonia Zerbesias, "APEC Squabble Could Cost CBC," *Toronto Star*, 20 October 1998.
7 Ibid.
8 Rosemary Speirs, "PM Gets a Rough Ride—and Shows a Weak Spot," *Toronto Star*, 12 December 1996.
9 *Toronto Star*, "The Prime Minister Is Lying," editorial, 12 December 1996.
10 Canadian Press, "Testy PM Lashes Out at Media—Chrétien Flays Sound-Bite Journalism," *Winnipeg Free Press*, 22 December 1996.
11 Edward Greenspon and Scott Feschuk, "PM Jocular about Self, Government Ministers Preoccupied with His Job Can Campaign 'from the Back Bench,'" *Globe and Mail*, 19 December 1997.
12 Zerbisias, "APEC Squabble Could Cost CBC."
13 Valerie Lawton, "APEC Coverage Was Fair, CBC Says: Broadcaster Strongly Rejects Accusations from PM's Office," *Toronto Star*, 7 November 1998.
14 *CBC/Radio-Canada Annual Report 1996–1997* (Ottawa: CBC/Radio-Canada, 1997), 12.

15 Rowland, *Canada Lives Here*, 46.

16 Ibid., 50.

17 Cass Sunstein, *#Republic: Divided Democracy in the Age of Social Media* (Princeton, NJ: Princeton University Press, 2017), 18–20.

18 Ibid., 141.

19 Elihu Katz, "And Deliver Us from Segmentation," *Annals of the American Academy of Political and Social Science* 546, no. 1 (July 1996): 22–33.

20 Alain Saulnier, *Losing Our Voice: Radio-Canada under Siege* (Toronto: Dundurn, 2015).

21 Chris Cobb, "CBC Television Review May Lead to Showdown: Committee MPs Hint Getting Consensus Is Struggle," *Edmonton Journal*, 19 February 2008.

22 Bruce Cheadle, "Heritage Minister Warns CBC Execs to Curb Expenses in Tough Times," Canadian Press, 19 November 2008.

23 Chris Cobb, "Most Voters Support Public Funding for CBC, Poll Finds: Group Wants Harper to Clarify Position on 'Commercialization' of CBC Radio," *Ottawa Citizen*, 2 October 2008.

24 Bruce Cheadle, "Tories Set for CBC Cuts, Position Critics Claim," Canadian Press, 21 November 2008.

25 Don Martin, "Minister Prefers Ad-Less CBC," *National Post*, 17 March 2009.

26 Jane Taber, "CBC Clears Pollster, Criticizes 'Paranoia-Tinged' Tories," *Globe and Mail*, 19 May 2010.

27 Vince Carlin, quoted in ibid.

28 Andrew Coyne, "A Real Tory Government Would Reform CBC," *Windsor Star*, 31 May 2014.

29 Ibid.

30 Ibid.

31 Jennifer Ditchburn, "Liberals Defend 'Total' Funding: Online Petition Recognizes 'Profound Importance' of Public Broadcasting Despite Conservative Efforts to Paint It as an 'Economic Scapegoat,'" Canadian Press, 13 October 2011.

32 Saulnier, *Losing Our Voice*, 137.

33 Ibid., 155–68.

34 *CBC/Radio-Canada Annual Report 2014–2015* (Ottawa: CBC/Radio-Canada, 2015).

35 Saulnier, *Losing Our Voice*, 178.

36 Ibid.

37 *Winnipeg Free Press*, "CBC CEO Disputes Harper Comment over Funding," 29 September 2015.

38 Friends of Canadian Broadcasting, "Bashing the CBC No Recipe for Tory Growth," news release, 24 May 2017.

39 Saulnier, *Losing Our Voice*, 176.

40 Ibid., 177–78.

41 Joe Chidley, "Juneau Reports on CBC," *Canadian Encyclopedia*, 17 March 2003, http://www.thecanadianencyclopedia.ca/en/article/juneau-reports-on-cbc/.

42 Ibid.

43 Guy Dixon, "Committee Reviews CBC Mandate," *Globe and Mail*, 10 March 2007, https://www.theglobeandmail.com/arts/committee-reviews-cbc-mandate/article17992503/.

44 Ibid.

45 Canada, Parliament, House of Commons, Standing Committee on Canadian Heritage, *CBC/Radio-Canada: Defining Distinctiveness in the Changing Media Landscape*, Report of the Standing Committee on Canadian Heritage, February 2008, http://publications.gc.ca/pub?id=9.545102&sl=0.

46 Canada, Parliament, Senate, Standing Senate Committee on Transport and Communications, "Final Report on the Canadian News Media," 2 vols., June 2006, https://sencanada.ca/content/sen/committee/391/tran/rep/repfinjun06vol1-e.htm#_Toc138058341.

47 Canada, Parliament, Senate, Standing Senate Committee on Transport and Communications, *Time for Change: The CBC/Radio-Canada in the Twenty-First Century*, Report of the Standing Committee on Transport and Communications, July 2015, https://sencanada.ca/Content/SEN/Committee/412/trcm/rep/rep14jul15-e.pdf.

48 Ibid., xi.

49 Canada, Parliament, Senate, "A Plan for a Vibrant and Sustainable CBC/Radio-Canada: Minority Report by Senator Art Eggleton in Response to the Standing Senate Committee on Transport and Communications Report on the CBC/Radio Canada," July 2015, https://senatorarteggleton.ca/wp-content/uploads/2018/11/A-Sustainable-CBC-Minority-Report-by-Senator-Art-Eggleton.pdf.

50 Carol Goar, "The Senate's Shoddy Report on CBC," *Toronto Star*, 26 July 2015.

51 Canada, Parliament, House of Commons, Standing Committee on Canadian Heritage, *Disruption: Change and Churning in Canada's Media Landscape*, Report of the Standing Committee on Canadian Heritage, June 2017, http://www.ourcommons.ca/DocumentViewer/en/42-1/CHPC/report-6/page-114#30.

52 Public Policy Forum, *The Shattered Mirror: News, Democracy and Trust in the Digital Age* (Ottawa: Public Policy Forum, January 2017), 94, https://shatteredmirror.ca/.

53 CBC/Radio-Canada, "Public Policy Forum Report: A Welcome Contribution to Discussions on the Future of Canadian Culture,"

news release, 26 January 2017, http://www.cbc.radio-canada.ca/en/media-centre/2017/01/26/.
54 Ibid.

4. The CBC in the Digital Storm

1 Collected from Google and YouTube.
2 Communic@tions Management Inc., "Netflix in Canada: Binge-Watching! Tax Issues! Regulatory Uncertainty!," discussion paper, 5 April 2018.
3 Derek Thompson, *Hit Makers: The Science of Popularity in an Age of Distraction* (New York: Penguin, 2017), 34.
4 Ibid., 243.
5 Josef Adalian, "Almost 500 Scripted Shows Aired in 2018, but We Still Haven't Hit Peak TV," Vulture.com, 14 December 2018.
6 Ibid.
7 Anita Elberse, *Blockbusters: Hit-Making, Risk-Taking and the Big Business of Entertainment* (New York: Henry Holt, 2013), 42.
8 Thompson, *Hit Makers*, 37.
9 Sam Alter, quoted in ibid., 131.
10 Markus Prior, *Post-broadcast Democracy* (Cambridge: Cambridge University Press, 2007).
11 Thompson, *Hit Makers*, 64.
12 Matthew Hindman, *The Internet Trap: How the Digital Economy Builds Monopolies and Undermines Democracy* (Princeton, NJ: Princeton University Press, 2018), 143.
13 Robert Putnam, *Bowling Alone: The Collapse and Revival of American Community* (New York: Simon & Schuster 2000), 231.
14 Steven Levitsky and Daniel Ziblatt, *How Democracies Die* (New York: Crown, 2018), 97.
15 Jürgen Habermas, *The Structural Transformation of the Public Sphere* (Cambridge, MA: MIT Press, 1989).
16 Michael Wolff, *Television Is the New Television* (New York: Penguin, 2015), 49.
17 Ibid., 57.
18 See Sherry Turkle, *Reclaiming Conversation* (New York: Penguin, 2015); see also Henry Jenkins, Mizuko Ito, and danah boyd, *Participatory Culture in a Networked Era: A Conversation on Youth, Learning, Commerce, and Politics* (Cambridge: Polity, 2016).
19 Florian Sauvageau, keynote address presented at the "How Canadians Communicate Politically: The Next Generation" conference, Banff, AB, October 2009.
20 Hindman, *The Internet Trap*, 149.
21 Ibid., 60.
22 Ibid., 2.

23 Ibid., 3.

24 Robert McChesney, *Digital Disconnect: How Capitalism Is Turning the Internet against Democracy* (New York: New Press, 2013), 210.

25 James Webster, *The Marketplace of Attention* (Cambridge, MA: MIT Press, 2014), 19.

26 José van Dijck, *The Culture of Creativity: A Critical History of Social Media* (New York: Oxford University Press, 2013).

27 Tim Wu, *The Master Switch: The Rise and Fall of Information Empires* (New York: Knopf, 2010), 235.

28 Elberse, *Blockbusters*, 69.

29 Communic@tions Management Inc., "Netflix in Canada"; see also Sophia Harris, "The Writing's on the Wall: Streaming Services Like Netflix Set to Overtake Cable TV," cbc.ca, 29 April 2018, https://www.cbc.ca/news/business/netflix-cable-tv-convergence-cord-cutting-1.4637065.

30 Communic@tions Management Inc., "Netflix in Canada."

31 Ibid.

32 Michael Smith and Rahul Telang, *Streaming, Sharing, Stealing: Big Data and the Future of Entertainment* (Cambridge, MA: MIT Press, 2016), chap. 1.

33 Jonah Weiner, "The Great Race to Rule Streaming TV," *New York Times Magazine*, 10 July 2019.

34 Andrew Romano, "The Way They Hook Us—for 13 Hours Straight," *Daily Beast*, May 2013, accessed 22 April 2018, http://www.andrewromano.net/40/the-way-they-hook-us-for-13-hours-straight.

35 Johanna Schneller, "A Dip into the Wide (Apple TV +) and Narrow (CBC's Gem) of Streaming Services," *Globe and Mail*, 1 April 2019, accessed 3 April 2019, https://www.theglobeandmail.com/arts/article-cbcs-gem-originals-are-pure-kind-canadiana/.

36 Statista.com, "Leading Social Networks Used Weekly for News in Canada as of February 2018," accessed 27 March 2019, https://www.statista.com/statistics/563514/social-networks-used-for-news-canada/.

37 Susan Delacourt, "Facebook Nation: Social Media Titan Dominates Canadian News," *iPolitics*, 2 February 2017, accessed 22 April 2018, https://ipolitics.ca/2017/02/07/facebook-nation-social-media-titan-now-dominates-canadian-news/.

38 Paige Cooper, "41 Facebook Stats That Matter to Marketers in 2019," *Hootsuite* (blog), accessed 28 March 2019.

39 John Gramlich, "10 Facts about Americans and Facebook," Pew Research Center, 10 April 2018, https://www.pewresearch.org/fact-tank/2019/05/16/facts-about-americans-and-facebook/.

40 Sherry Turkle, *Reclaiming Conversation: The Power of Talk in a Digital Age* (New York: Penguin, 2015), 126.

41 Bernard Harcourt, *Exposed: Desire and Disobedience in the Digital Age* (Cambridge, MA: Harvard University Press, 2015).

42 Ibid., chap. 4.

43 Scott Galloway, *The Four: The Hidden DNA of Amazon, Apple, Facebook, and Google* (New York: Penguin, 2017), 103–04.

44 Cass Sunstein, *#Republic: Divided Democracy in the Age of Social Media* (Princeton, NJ: Princeton University Press, 2017), 86.

45 Eli Pariser, *The Filter Bubble: What the Internet Is Hiding from You* (New York: Penguin, 2011), 47.

46 boyd, in Jenkins, Ito, and boyd, *Participatory Culture*, 28.

47 Daniel Kreiss, "The Media Are about Identity, Not Information," in *Trump and the Media*, ed. Pablo Boczkowski and Zizi Papacharissi (Cambridge, MA: MIT Press, 2018), 93–99.

48 Hindman, *The Internet Trap*, 71.

49 Elizabeth Dubois and Grant Blank, "The Echo Chamber Is Overstated: The Moderating Effect of·Political Interest and Diverse Media," *Information, Communication & Society* 21, no. 5 (2018): 729–45, https://doi.org/10.1080/1369118X.2018.1428656.

50 Sunstein, *#Republic*, 26–27.

51 Yascha Mounk, *The People vs. Democracy: Why Our Freedom Is in Danger and How to Save It* (Boston, MA: Harvard University Press, 2018).

52 Webster, *Marketplace of Attention*, 35.

53 Kate Losse, "The Real 'News Curators' at Facebook Are the Engineers Who Write Its Algorithms," *Real Future*, 5 October 2016, accessed 10 May 2018, https://splinternews.com/the-real-news-curators-at-facebook-are-the-engineers-wh-1793856718.

54 Franklin Foer, *The World without Mind: The Existential Threat of Big Tech* (New York: Penguin, 2017), 140.

55 Ibid., 148–49.

5. The Collapse of Sports and News

1 David Shoalts, *Hockey Fight in Canada* (Madeira Park, BC: Douglas & McIntyre, 2018).

2 Ibid., 132–33.

3 Ibid., 8, 65–66, 91.

4 Peter Gzowski, quoted in Jason Blake, *Canadian Hockey Literature* (Toronto: University of Toronto Press, 2010), 27.

5 Ibid.

6 David Shoalts, "Top Hockey Producer's Firing before NHL Playoffs Hints at Rogers Turmoil," *Globe and Mail*, 7 April 2016, https://www.theglobeandmail.com/sports/hockey/top-hockey-producers-firing-before-nhl-playoffs-hints-at-rogers-turmoil/article29548827/.

7 Shoalts, *Hockey Fight in Canada*, 102.

8 cbc.ca, "Rogers, CBC Ink Deal to Keep Hockey Night in Canada on CBC until 2026, Streaming Included," 19 December 2017, http://www.cbc.ca/news/business/cbc-rogers-hockey-deal-1.4456017.

9 Scott Stinson, "CBC's Strange Partnership with Rogers to Show NHL Games Remains at Odds with Identity as Public Broadcaster," *National Post*, 25 May 2017, http://news.nationalpost.com/sports/nhl/cbcs-strange-partnership-with-rogers-to-show-nhl-games-remains-at-odds-with-identity-as-public-broadcaster.

10 Communic@tions Management Inc., "Betting on the Games: Canadian Television's Spending on Sports Programming Now Exceeds $1 Billion a Year," discussion paper, 14 March 2019.

11 Christopher Waddell, "The Hall of Mirrors," in *How Canadians Communicate V: Sports*, ed. David Taras and Christopher Waddell (Edmonton: Athabasca University Press, 2016), 46, http://www.aupress.ca/books/120244/ebook/03_Taras_Waddell_2016-How_Canadians_Communicate_V_Sports.pdf.

12 Ibid., 53.

13 Ibid.

14 Wikipedia, "List of Most Watched Television Broadcasts in Canada," accessed 15 August 2018, https://en.wikipedia.org/wiki/List_of_most_watched_television_broadcasts_in_Canada.

15 Richard Yao, "The Unraveling of Live Sports TV," IPG Media Lab, 22 February 2018.

16 Ibid.

17 Susan Jacoby, *Why Baseball Matters* (New Haven: CT: Yale University Press, 2018), 66.

18 Canada, Parliament, House of Commons, Standing Committee on Canadian Heritage, *Evidence*, 42nd Parliament, 1st Session, No. 032, 25 October 2016, http://www.ourcommons.ca/DocumentViewer/en/42-1/CHPC/meeting-32/evidence.

19 MoffettNathanson, quoted in Peter Kafka, "The NFL Can't Blame Trump Anymore: It Is Facing a 'Structural Decline in Viewership,'" *recode*, 29 January 2018.

20 Derek Thompson, "Why NFL Ratings Are Plummeting: A Two-Part Theory," *Atlantic*, 1 February 2018; see also Frank Pallotta, "NFL Ratings Rebound after Two Seasons of Declining Viewership," CNN Business, 3 January 2019.

21 Yao, "The Unraveling of Live Sports TV."

22 Carson Kessler, "MLB Attendance Drops to Lowest Average in 15 Years," *Fortune*, 15 June 2018.

23 Associated Press, "World Series 4th—Least-Watched, Averaging 14.1M Viewers," AP news.com, 30 October 2018.

24 Bobby Burack, "Why Aren't NBA's Ratings Declines Being Discussed in Media Like the NFL's Were?," biglead.com, 18 January 2019, https://thebiglead.com/2019/01/18/nba-ratings-decline-nfl/.

25 Statista.com, "Average TV Viewership of NBA Finals in the United
    States from 2002 to 2018," https://www.statista.com/statistics/240381/
    nba-finals-tv-ratings-in-the-united-states/.

26 James Bradshaw, "Two Years into Its $5.2-Billion NHL Deal, Did Rogers
    Make the Right Call?," *Globe and Mail*, 27 May 2016, http://www.the
    globeandmail.com/report-on-business/changes-in-nhl-fan-support-
    have-broadcasters-taking-the-hit/article30194300/.

27 Bill Brioux, "NUMBERS IN CANADA: Raptors Crack 3M, Set B-ball
    Benchmark," brioux.tv, 27 May 2019, https://brioux.tv/blog/2019/05/27/
    numbers-in-canada-raptors-crack-3m-set-b-ball-benchmark/.

28 Communic@tions Management Inc., "Betting on the Games," 12.

29 Ibid.

30 Dave Lozo, "The NHL Has a Problem: All Its Fans Are Bloody Old," Vice
    Sports, 8 June 2017.

31 Sports Media Watch, "National Hockey League Stanley Cup Games
    and TV Viewership from 2008–2018 (in Millions)," accessed 16 August
    2018, https://www.statista.com/statistics/305818/average-tv-ratings-
    nhl-stanley-cup-games/.

32 Victoria Ahern, "Canadian Olympics Junkies Keeping Odd Hours to Catch
    Live Events," Canadian Press, 13 February 2018, https://globalnews.ca/
    news/4022413/winter-olympics-2018-viewing-junkies/.

33 Richard Sandomir, "With Audiences Shrinking, NBC Looks Cautiously
    to Olympics in Asia," *New York Times*, 22 August 2016, http://www.
    nytimes.com/2016/08/23/sports/with-audience-shrinking-nbc-looks-
    cautiously-to-olympics-in-asia.html?smprod=nytcore-iphone&smid=
    nytcore-iphone-s.

34 Jamie Strachin, "Cable TV Cord Cutters Threaten Sports Business," cbc.ca,
    13 May 2017, http://www.cbc.ca/sports/cable-sports-tv-cord-cutting-
    1.4112895.

35 Clay Travis, "ESPN Loses 500,000 Subscribers in April, Outkick the
    Coverage," 1 May 2018, https://www.outkickthecoverage.com/espn-
    loses-500000-subscribers-april/.

36 Yao, "The Unraveling of Live Sports TV."

37 Susan Jacoby, *Why Baseball Matters*.

38 Ibid., 66.

39 Ibid., 26.

40 Canadian Radio-television and Telecommunications Commission (CRTC),
    Communications Monitoring Report 2017 (Ottawa: CRTC, 2018), table 4.2.23.

41 CBC/Radio-Canada, response to Access to Information request from
    Patrick McCurdy, 18 April 2019.

42 Communic@tions Management Inc., "An Assessment of Federal Government
    Programs Impacting on Local Journalism in Canada," 24 April 2019, 10.

43 Barry Kiefl, "Has Canada Lost Faith in the CBC?," *Huffington Post*, 28 July 2014, http://www.huffingtonpost.ca/barry-kiefl/cbc-public-opinion_b_5404479.html.

44 Adam Alter, *Irresistible: The Rise of Addictive Technology and the Business of Keeping Us Hooked* (New York: Penguin, 2017), 15.

45 John Doyle, "It's about Time: We've Put Up with Mansbridge and His Pompous Ilk for Too Long," *Globe and Mail*, 6 September 2016.

46 John Doyle, quoted in Wade Rowland, *Canada Lives Here: The Case for Public Broadcasting* (Westmount, QC: Linda Leith Publishing, 2015), 95.

47 Ibid.

48 John Doyle, "Revamped The National Is a Harebrained Muddle," *Globe and Mail*, 10 November 2017.

49 Bill Brioux, "New *National* Ratings No Better Than Mansbridge's, but CBC Isn't Worried," *Toronto Star*, 16 November 2017, https://www.thestar.com/entertainment/television/2017/11/16/new-national-ratings-no-better-than-mansbridges-but-cbc-isnt-worried.html.

50 Bill Brioux, "NUMBERS ACROSS CANADA: Final Verdict Reached on Street Legal," 11 April 2019, https://brioux.tv/2019/04/numbers-across-canada-final-verdict-reached-on-street-legal/.

51 Ken Auletta, *Frenemies: The Epic Disruption of the Ad Business (and Everything Else)* (New York: Penguin, 2018), 262.

52 Franklin Foer, *World without Mind* (New York: Penguin, 2017), 146.

53 Cass Sunstein, *#Republic: Divided Democracy in the Age of Social Media* (Princeton, NJ: Princeton University Press, 2017).

54 Pablo J. Boczkowski and Seth Lewis, "The Center of the Universe No More: From Self-Centered Stance of the Past to the Relational Mindset of the Future," in *Trump and the Media*, ed. Pablo J. Boczkowski and Zizi Papacharissi (Cambridge: MA: MIT Press, 2018), 177–85.

55 British Broadcasting Corporation, "BBC News Announces New Investments in Canada," news release, 30 June 2016, http://www.bbc.co.uk/corporate2/mediacentre/worldnews/2016/bbc-news-announces-new-investments-in-canada.

6. The Trials and Triumphs of the CBC's Online World

1 Michael Wolfe, *Television Is the New Television: The Unexpected Triumph of Old Media in the Digital Age* (New York: Portfolio, 2015), 49.

2 Matthew Hindman, *The Internet Trap: How the Digital Economy Builds Monopolies and Undermines Democracy* (Princeton, NJ: Princeton University Press, 2018), 108.

3 Ibid., 106.

4 Ibid., 107.

5 First CBC producer, correspondence with authors, 5 April 2019.

6 CBC journalist, correspondence with authors, 5 April 2019.

7 Ibid.

8 Second CBC producer, correspondence with authors, 4 April 2019.

9 Barry Kiefl, "Cutting Government Appointees Could Stall CBC's Downwards Spiral," *Huffington Post*, 19 September 2016, http://www.huffingtonpost.ca/barry-kiefl/cbc-problems_b_12049472.html.

10 Gregory Taylor, quoted in Tom Jokinen, "What Is the CBC Good For?," *Walrus*, September 2017, 20–26.

11 John Doyle, "Conventional Television Is Doing Just Fine, Thanks for Asking," *Globe and Mail*, 31 August 2017, https://beta.theglobeandmail.com/arts/television/john-doyle-conventional-television-is-doing-just-fine-thanks-for-asking/article36138342/.

12 Communic@tions Management Inc., "Supporting Canadian Journalism: An Assessment of Selected Federal Government Programs," research note, 24 April 2019, 10.

13 Ross Winn, "2019 Podcast Stats & Facts," 6 March 2019, https://www.podcastinsights.com/podcast-statistics/.

14 Genna Buck, "Why Canadian Podcasters Are Being Drowned Out by American Offerings," *Metronews.ca*, 29 July 2016.

15 Brad Clark and Archie McLean, "Revenge of the Nerds: How Public Radio Has Transformed Listening to Audio, and Come to Dominate Podcasting," research paper, November 2017 (Calgary: Mount Royal University).

16 Erica Ngao, "The Podcast Evolution," *Ryerson Review of Journalism* 43, no. 1 (Spring 2017): 30–37.

7. More Dashed Hopes

1 Barry Kiefl, "CBC Must Abandon Ads and Find New Sources of Funding," *Huffington Post*, 10 February 2016, https://www.huffingtonpost.ca/barry-kiefl/cbc-advertising-revenues_b_9201924.html.

2 Canada, Parliament, House of Commons, Standing Committee on Canadian Heritage, Hubert T. Lacroix testimony, 42nd Parliament, 1st Session, Meeting No. 32, 25 October 2016, http://www.parl.gc.ca/HousePublications/Publication.aspx?Language=e&Mode=1&Parl=42&Ses=1&DocId=8538787.

3 CBC/Radio-Canada, "Letter to the Standing Committee on Canadian Heritage: Limiting Access to the Digital Public Space Is Not in the Public Interest," letter from Hubert T. Lacroix, 21 November 2016, https://cbc.radio-canada.ca/en/media-centre/2016/november/letter-limiting-access-digital-public-space-not-in-public-interest.

4 *CBC/Radio-Canada Annual Report 2017–2018* (Ottawa: CBC/Radio-Canada, 2018), 102.

5 CBC/Radio-Canada, *A Creative Canada: Strengthening Canadian Content in a Digital World*, submission in support of the government's public consultation on the future of Canadian content in a digital world, http://future.cbc.ca/images/acreativecanada.pdf.

6 Ibid., 27.

7 Ibid., 29.

8 Kate Taylor, "Ad-Free CBC Could Serve as Rallying Point for Canadian Creativity," *Globe and Mail*, 3 December 2016, https://beta.theglobeandmail.com/arts/television/ad-free-cbc-could-act-as-a-rallying-point-for-canadian-creativity/article33138985/?ref=http://www.theglobeandmail.com.

9 Simon Houpt and Susan Krashinsky, "CBC Ad-Free Proposal Shakes Up Heritage Discussion," *Globe and Mail*, 2 December 2016, http://www.theglobeandmail.com/report-on-business/cbcs-ad-free-proposal-shakes-up-heritage-discussion/article33186433/.

10 Ibid.

11 Jeffrey Dvorkin, quoted in ibid.

12 John Doyle, "Does the CBC Define Canadian Culture?," *Globe and Mail*, 28 May 2017, https://www.theglobeandmail.com/arts/television/john-doyle-does-the-cbc-define-canadian-culture/article35119806/.

13 Richard Stursberg, "The CBC Is Dying. Here's How to Save It," *Toronto Star*, 19 November 2015, http://www.thestar.com/opinion/commentary/2015/11/19/the-cbc-is-dying-heres-how-to-save-it.html.

14 Robert Everett-Green, "Mélanie Joly Urges Patience on Long Road toward Cultural Policy Shakeup," *Globe and Mail*, 28 April 2017, https://www.theglobeandmail.com/news/politics/melanie-joly-bids-patience-as-she-marches-toward-cultural-policy-shakeup/article34854385/.

15 John Doyle, "The Netflix Deal Is a Very Sweet Deal for Netflix, Not Canada," *Globe and Mail*, 30 September 2016, https://beta.theglobeandmail.com/arts/television/the-netflix-deal-is-a-very-sweet-deal-for-netflix-not-canada/article36421246/.

16 Ian Morrison, quoted in Alex Ballingall, "$500-Million Netflix Deal Anchors Canada's Culture Plan," *Toronto Star*, 28 September 2017, https://www.thestar.com/news/canada/2017/09/28/500-million-netflix-deal-anchors-canadas-culture-plan.html.

17 Richard Stursberg with Stephen Armstrong, *The Tangled Garden: A Canadian Cultural Manifesto for the Digital Age* (Toronto: James Lorimer, 2019), 176–82.

18 Amanda Lotz, *We Now Disrupt This Broadcast* (Boston, MA: MIT Press, 2018).

19 Canada, Department of Finance, "Equality + Growth: A Strong Middle Class," 2018 Federal Budget, 27 February 2018, 183–84.

20 Canada, Department of Finance, "2018 Fall Economic Statement: Investing in Middle Class Jobs," 21 November 2018, 40–41.

21 Christopher Waddell, "Government Funding for Journalism: To What End?," *The Conversation*, 21 March 2019.

22 Ibid.

23 Anita Elberse, *Blockbusters: Hit-Making, Risk-Taking and the Big Business of Entertainment* (New York: Henry Holt, 2013), chap. 1.

24 Lotz, *We Now Disrupt This Broadcast*, x.

25 Susan Krashinsky Robertson and Laura Stone, "Liberal Deal Lets Netflix Play by Its Own Rules in Canadian Broadcasting," *Globe and Mail*, 30 September 2017, https://beta.theglobeandmail.com/report-on-business/liberal-deal-puts-netflix-on-its-own-plane-in-canadian-broadcasting/article36449302/.

26 James Bradshaw and Daniel Leblanc, "Canadian Experts Unite for Cultural Policy Advisory Group," *Globe and Mail*, 28 June 2016, http://www.theglobeandmail.com/report-on-business/canadian-experts-unite-for-cultural-policy-advisory-group/article30635179/.

27 Canada, Department of Canadian Heritage, "Government of Canada Launches Review of Telecommunications and Broadcasting Acts," 5 June 2018, https://www.canada.ca/en/canadian-heritage/news/2018/06/government-of-canada-launches-review-of-telecommunications-and-broadcasting-acts.html.

28 Chantal Hébert, "Ottawa's Netflix Deal Angers Culture-Conscious Quebec: Hébert," *Toronto Star*, 29 September 2016, https://www.thestar.com/amp/news/canada/2017/09/29/ottawas-netflix-deal-angers-culture-conscious-quebec.html.

29 Robert Everett-Green, "A Stunning Fall from Grace for Mélanie Joly," *Globe and Mail*, 4 October 2017, https://www.theglobeandmail.com/opinion/a-stunning-fall-from-grace-for-melanie-joly/article36497712/.

30 Simon Houpt, "Ex-journalist Tom Clark, Actor Colin Feore among CBC Board Advisory Team Unveiled by Joly," *Globe and Mail*, 20 June 2017, https://www.theglobeandmail.com/arts/television/heritage-minister-melanie-joly-unveils-cbc-board-advisory-team/article35404032/.

31 Markus Prior, *Post-broadcast Democracy* (New York: Cambridge University Press, 2007).

8. Reinvent the CBC or Allow It to Die

1 Catherine Tait, "CBC/Radio-Canada: At the Heart of Canadians' Lives," speech delivered to the Montreal Chamber of Commerce,

3 May 2019, https://cbc.radio-canada.ca/en/media-centre/2019/speaking-notes-catherine-tait-chamber-commerce-mtl.

2 Stursberg with Armstrong, *The Tangled Garden*, 131.

3 Communic@tions Management Inc., "Finding the News 2.0: How Age, Language, and Geography Influence Canadians' Media Choices," discussion paper, 30 October 2018.

4 CBC/Radio-Canada, "Your Stories, Taken to Heart," the CBC's three-year strategy, 22 May 2019, https://cbc.radio-canada.ca/en/vision/strategy/your-stories-taken-to-heart.

5 Ibid.

6 Ibid.

7 Ibid.

8 Catherine Tait, "CBC/Radio-Canada."

9 Matthew Hindman, *The Internet Trap: How the Digital Economy Builds Monopolies and Undermines Democracy* (Princeton, NJ: Princeton University Press, 2018), 37.

10 Hayden Watters, "Vote Compass Has Been Used 365K Times: What Has Been Learned?," cbc.ca, 6 June 2018, https://www.cbc.ca/news/canada/toronto/live-blog-vote-compass-1.4694038.

Bibliography

Adalian, Josef. "Almost 500 Scripted Shows Aired in 2018, but We Still Haven't Hit Peak TV." Vulture.com, 14 December 2018. https://www.vulture.com/2018/12/peak-tv-scripted-originals-2018.html.

Ahern, Victoria. "Canadian Olympics Junkies Keeping Odd Hours to Catch Live Events." Canadian Press, 13 February 2018. https://globalnews.ca/news/4022413/winter-olympics-2018-viewing-junkies/.

Alter, Adam. *Irresistible: The Rise of Addictive Technology and the Business of Keeping Us Hooked.* New York: Penguin, 2017.

Anderson, Benedict. *Imagined Communities: Reflections on the Origins and Spread of Nationalism.* London: Verso, 1983.

Arriaza Ibarra, Karen, Eva Nowak, and Raymond Kuhn, eds. *Public Service Media in Europe: A Comparative Approach.* London: Routledge, 2015.

Arterton, F. Christopher. "The Media Politics of Presidential Campaigns." In *The Race for the Presidency: The Media and the Nominating Process,* edited by James David Barber, 3–25. Englewood Cliffs, NJ: Prentice Hall, 1978.

Associated Press. "World Series 4th—Least-Watched, Averaging 14.1M Viewers." AP news.com, 30 October 2018.

Auletta, Ken. *Frenemies: The Epic Disruption of the Ad Business (and Everything Else).* New York: Penguin, 2018.

Ballingall, Alex. "$500-million Netflix Deal Anchors Canada's Culture Plan." thestar.com, 28 September 2017. https://www.thestar.com/news/canada/2017/09/28/500-million-netflix-deal-anchors-canadas-culture-plan.html.

Bank of Canada. Inflation calculator. https://www.bankofcanada.ca/rates/related/inflation-calculator/.

Barber, James David. *The Race for the Presidency.* Englewood Cliffs, NJ: Prentice Hall, 1978.

Bastien, Frederick. *Breaking News: Politics, Journalism and Infotainment on Québec Television.* Vancouver: UBC Press, 2018.

Blake, Jason. *Canadian Hockey Literature*. Toronto: University of Toronto Press, 2010.

Blatchford, Christie. "Mainstream Media Is Starving—but Certainly not the CBC." *National Post* online, 30 June 2018. https://nationalpost.com/opinion/christie-blatchford-mainstream-media-is-starving-but-certainly-not-the-cbc.

Boczkowski, Pablo, and Seth Lewis. "The Center of the Universe No More: From Self-Centered Stance of the Past to the Relational Mindset of the Future." In *Trump and the Media*, edited by Pablo Bocskowski and Zizi Papacharissi, 177–85. Cambridge: MA: MIT Press, 2018.

Bocskowski, Pablo, and Zizi Papacharissi, eds. *Trump and the Media*. Cambridge: MA: MIT Press, 2018, 1–6.

Bradshaw, James. "Two Years into Its $5.2-Billion NHL Deal, Did Rogers Make the Right Call?" *Globe and Mail*, 27 May 2016. http://www.theglobe andmail.com/report-on-business/changes-in-nhl-fan-support-have-broad casters-taking-the-hit/article30194300/.

Bradshaw, James, and Daniel Leblanc. "Canadian Experts Unite for Cultural Policy Advisory Group." *Globe and Mail*, 28 June 2016. http://www.theglobeandmail.com/report-on-business/canadian-experts-unite-for-cultural-policy-advisory-group/article30635179/.

Brioux, Bill. "New National Ratings No Better than Mansbridge's, but CBC Isn't Worried." *Toronto Star*, 16 November 2017. https://www.thestar.com/entertainment/television/2017/11/16/new-national-ratings-no-better-than-mansbridges-but-cbc-isnt-worried.html 1/4.

– "NUMBERS ACROSS CANADA: Final Verdict Reached on Street Legal." brioux.tv, 11 April 2019. https://brioux.tv/2019/04/numbers-across-canada-final-verdict-reached-on-street-legal/.

– "NUMBERS IN CANADA: Raptors Crack 3M, Set B-ball Benchmark." brioux.tv, 27 May 2019. https://brioux.tv/blog/2019/05/27/numbers-in-canada-raptors-crack-3m-set-b-ball-benchmark/.

British Broadcasting Corporation. "BBC News Announces New Investments in Canada." News release. 30 June 2016. http://www.bbc.co.uk/corporate2/mediacentre/worldnews/2016/bbc-news-announces-new-investments-in-canada.

Buck, Genna. "Why Canadian Podcasters Are Being Drowned Out by American Offerings." *Metronews.ca*, 29 July 2016.

Burack, Bobby. "Why Aren't NBA's Ratings Declines Being Discussed in Media Like the NFL's Were?" biglead.com, 18 January 2019. https://thebiglead.com/2019/01/18/nba-ratings-decline-nfl/.

Canada. Department of Canadian Heritage. "Government of Canada Launches Review of Telecommunications and Broadcasting Acts."

News release. 5 June 2018. https://www.canada.ca/en/canadian-heritage/news/2018/06/government-of-canada-launches-review-of-telecommunications-and-broadcasting-acts.htm.

Canada. Department of Finance. "Equality + Growth: A Strong Middle Class." 2018 Federal Budget. 27 February 2018.

– "Investing in Middle Class Jobs." News release. 21 November 2018.

Canada. Parliament. House of Commons. Standing Committee on Canadian Heritage. *CBC/Radio-Canada: Defining Distinctiveness in the Changing Media Landscape.* Report of the Standing Committee on Canadian Heritage. February 2008. http://www.ourcommons.ca/DocumentViewer/en/39-2/CHPC/report-6\.

– *Disruption: Change and Churning in Canada's Media Landscape.* Report of the Standing Committee on Canadian Heritage. June 2017. http://www.ourcommons.ca/DocumentViewer/en/42-1/CHPC/report-6/page-114#30.

– *Evidence.* Hubert T. Lacroix testimony. 42nd Parliament, 1st Session, No. 032. 25 October 2016. http://www.ourcommons.ca/DocumentViewer/en/42-1/CHPC/meeting-32/evidence.

Canada. Parliament. Senate. Standing Senate Committee on Transport and Communications. *CBC/Radio-Canada: Defining Distinctiveness in the Changing Media Landscape.* Report of the Standing Committee on Canadian Heritage, February 2008, http://publications.gc.ca/pub?id=9.545102&sl=0.

– "Final Report on the Canadian News Media." 2 vols. June 2006. https://sencanada.ca/content/sen/committee/391/tran/rep/repfinjun06vol1-e.htm.

– *Time for Change: The CBC/Radio-Canada in the Twenty-First Century.* Report of the Standing Senate Committee on Transport and Communications. July 2015. https://sencanada.ca/Content/SEN/Committee/412/trcm/rep/rep14jul15-e.pdf.

Canadian Press. "CBC CEO Disputes Harper Comment over Funding." 29 September 2015. https://www.theglobeandmail.com/news/politics/cbc-ceo-disputes-harper-comment-over-funding/article26588069/.

– "Testy PM Lashes Out at Media: Chrétien Flays Sound-Bite Journalism." *Winnipeg Free Press*, 22 December 1996.

Canadian Radio-television and Telecommunications Commission (CRTC). *Communications Monitoring Report 2017.* Ottawa: CRTC, 2018.

– *Communications Monitoring Report 2018.* Ottawa: CRTC, 2019.

cbc.ca. "Rogers, CBC Ink Deal to Keep Hockey Night in Canada on CBC until 2026, Streaming Included." 19 December 2017. http://www.cbc.ca/news/business/cbc-rogers-hockey-deal-1.4456017.

CBC/Radio-Canada. *Annual Report 1996–1997.* Ottawa: CBC/Radio-Canada, 1997.

– *Annual Report 2014–2015*. Ottawa: CBC/Radio-Canada, 2015.

– *Annual Report 2017–2018*. Ottawa: CBC/Radio-Canada, 2018.

– *A Creative Canada: Strengthening Canadian Content in a Digital World*. Submission in support of the government's public consultation on the future of Canadian content in a digital world. Ottawa: CBC/Radio-Canada, 28 November 2016. http://future.cbc.ca/images/acreative canada.pdf.

– "Letter to the Standing Committee on Canadian Heritage: Limiting Access to the Digital Public Space Is Not in the Public Interest." Letter from Hubert T. Lacroix, 21 November 2016, https://cbc.radio-canada.ca/en/media-centre/2016/november/letter-limiting-access-digital-public-space-not-in-public-interest.

– "Public Policy Forum Report—A Welcome Contribution to Discussions on the Future of Canadian Culture." News release. 26 January 2017. http://www.cbc.radio-canada.ca/en/media-centre/2017/01/26/.

– Response to Access to Information request from Patrick McCurdy. 18 April 2019.

– "Thank You Canada! CBC/Radio-Canada Doubles Its Digital Reach Two and a Half Years Early." News release. 11 December 2017. https://www.newswire.ca/news-releases/thank-you-canada-cbcradio-canada-doubles-its-digital-reach-two-and-a-half-years-early-663415263.htm.

– "Your Stories, Taken to Heart." New Three-Year Plan. 22 May 2019. https://cbc.radio-canada.ca/en/vision/strategy/your-stories-taken-to-heart.

Cheadle, Bruce. "Heritage Minister Warns CBC Execs to Curb Expenses in Tough Times." Canadian Press, 19 November 2008.

– "Tories Set for CBC Cuts, Position Critics Claim." Canadian Press, 21 November 2008.

Chidley, Joe. "Juneau Reports on CBC." *Canadian Encyclopedia*. 17 March 2003. http://www.thecanadianencyclopedia.ca/en/article/juneau-reports-on-cbc/.

Clark, Brad, and Archie McLean. "Revenge of the Nerds: How Public Radio Has Transformed Listening to Audio, and Come to Dominate Podcasting." Research paper. November 2017. Calgary: Mount Royal University.

Cobb, Chris. "CBC Television Review May Lead to Showdown: Committee MPs Hint Getting Consensus Is Struggle." *Edmonton Journal*, 19 February 2008.

– "Most Voters Support Public Funding for CBC, Poll Finds: Group Wants Harper to Clarify Position on 'Commercialization' of CBC Radio." *Ottawa Citizen*, 2 October 2008.

– "Saul Enters CBC Debate." *National Post*, 30 January 2001.

Communic@tions Management Inc. "Betting on the Games: Canadian Television's Annual Spending on Sports Programming Now Exceeds $1 Billion a Year." Discussion paper. 14 March 2019.

– "Canada's Digital Divides." Discussion paper. 20 August 2015.
– "Finding the News 2.0: How Age, Language, and Geography Influence Canadians' Media Choices." Discussion paper. 30 October 2018.
– "Netflix in Canada: Binge-Watching! Tax Issues! Regulatory Uncertainty!" Discussion paper. 5 April 2018.
– "Supporting Canadian Journalism: An Assessment of Selected Federal Government Programs." Research note. 24 April 2019.
Cooper, Paige. "41 Facebook Stats That Matter to Marketers in 2019." *Hootsuite* (blog). Accessed 28 March 2019.
Cormack, Patricia, and James F. Cosgrave. *Desiring Canada: CBC Contests, Hockey Violence, and Other Stately Pleasures*. Toronto: University of Toronto Press, 2012.
Coyne, Andrew. "A Real Tory Government Would Reform CBC." *Windsor Star*, 31 May 2014. https://o.canada.com/news/national/coyne-a-real-conservative-government-would-reform-the-cbc.
DeCillia, Brooks, and Patrick McCurdy. "The Sound of Silence: The Absence of Public Service Values in Canadian Media Discourse about the CBC." *Canadian Journal of Communication* 41, no. 4 (2016): 547–67.
– "Viewing the CBC as a Public Good." *Policy Options* (24 November 2016). https://policyoptions.irpp.org/magazines/november-2016/viewing-the-cbc-as-a-public-good/.
Delacourt, Susan. "Facebook Nation: Social Media Titan Dominates Canadian News." *iPolitics*, 2 February 2017. Accessed 22 April 2018. https://ipolitics.ca/2017/02/07/facebook-nation-social-media-titan-now-dominates-canadian-news/.
Ditchburn, Jennifer. "Liberals Defend 'Total' Funding: Online Petition Recognizes 'Profound Importance' of Public Broadcasting Despite Conservative Efforts to Paint It as an 'Economic Scapegoat.'" Canadian Press, 13 October 2011.
Dixon, Guy. "Committee Reviews CBC Mandate." *Globe and Mail*, 10 March 2007. https://www.theglobeandmail.com/arts/committee-reviews-cbc-mandate/article17992503/.
Doyle, John. "Conventional Television Is Doing Just Fine, Thanks for Asking." *Globe and Mail*, 31 August 2017. https://beta.theglobeandmail.com/arts/television/john-doyle-conventional-television-is-doing-just-fine-thanks-for-asking/article36138342/.
– "Does the CBC Define Canadian Culture?" *Globe and Mail*, 28 May 2017. https://www.theglobeandmail.com/arts/television/john-doyle-does-the-cbc-define-canadian-culture/article35119806/.
– "It's about Time: We've Put Up with Mansbridge and His Pompous Ilk for Too Long." *Globe and Mail*, 6 September 2016. https://www.theglobeandmail.com/arts/television/its-about-time-weve-put-up-with-mansbridge-and-his-pompous-ilk-for-too-long/article31720560/.

– "The Netflix Deal Is a Very Sweet One for Netflix, Not Canada." *Globe and Mail*, 30 September 2016. https://beta.theglobeandmail.com/arts/television/the-netflix-deal-is-a-very-sweet-deal-for-netflix-not-canada/article36421246/.

– "Revamped The National Is a Harebrained Muddle," *Globe and Mail*, 10 November 2017. https://www.theglobeandmail.com/arts/television/revamped-the-national-is-a-harebrained-muddle/article36907586/.

Dubois, Elizabeth, and Grant Blank. "The Echo Chamber Is Overstated: The Moderating Effect of Political Interest and Diverse Media." *Information, Communication & Society* 21, no. 5 (2019): 729–45. https://doi.org/10.1080/1369118X.2018.1428656.

Elberse, Anita. *Blockbusters: Hit-Making, Risk-Taking and the Big Business of Entertainment*. New York: Henry Holt, 2013.

Everett-Green, Robert. "A Stunning Fall from Grace for Mélanie Joly." *Globe and Mail*, 4 October 2017. https://www.theglobeandmail.com/opinion/a-stunning-fall-from-grace-for-melanie-joly/article36497712/.

– "Mélanie Joly Urges Patience on Long Road toward Cultural Policy Shakeup." *Globe and Mail*, 28 April 2017. https://www.theglobeandmail.com/news/politics/melanie-joly-bids-patience-as-she-marches-toward-cultural-policy-shakeup/article34854385/.

Flaherty, David, and William McKercher, eds. *Southern Exposure: Canadian Perspectives on the United States*. Toronto: McGraw Hill Ryerson, 1982.

Foer, Franklin. *The World without Mind: The Existential Threat of Big Tech*. New York: Penguin, 2017.

Friends of Canadian Broadcasting. "Bashing the CBC No Recipe for Tory Growth." News release. 24 May 2017.

Galloway, Scott. *The Four: The Hidden DNA of Amazon, Apple, Facebook, and Google*. New York: Penguin, 2017.

Goar, Carol. "The Senate's Shoddy Report on CBC." *Toronto Star*, 26 July 2015.

Gramlich, John. "10 Facts about Americans and Facebook." Pew Research Center. 10 April 2018. https://www.pewresearch.org/fact-tank/2019/05/16/facts-about-americans-and-facebook/.

Greenspon, Edward, and Scott Feschuk. "PM Jocular about Self, Government Ministers Preoccupied with His Job Can Campaign from the Back Bench." *Globe and Mail*, 19 December 1997.

Grieco, Elizabeth. "Newsroom Employment Dropped Nearly a Quarter in Less Than 10 Years, with Greatest Decline at Newspapers." Pew Research Center. 30 July 2018. https://www.pewresearch.org/fact-tank/2018/07/30/newsroom-employment-dropped-nearly-a-quarter-in-less-than-10-years-with-greatest-decline-at-newspapers/.

Grossman, Michael, and Martha Kumar. *Portraying the President: The White House and the News Media*. Baltimore: Johns Hopkins University Press, 1981.

Habermas, Jürgen. *The Structural Transformation of the Public Sphere*. Cambridge, MA: MIT Press, 1989.

Harcourt, Bernard. *Exposed: Desire and Disobedience in the Digital Age.* Cambridge, MA: Harvard University Press, 2015.

Harris, Sophia. "The Writing's on the Wall: Streaming Services like Netflix Set to Overtake Cable TV." cbc.ca, 29 April 2018. https://www.cbc.ca/news/business/netflix-cable-tv-convergence-cord-cutting-1.4637065.

Hébert, Chantal. "Ottawa's Netflix Deal Angers Culture-Conscious Quebec: Hébert." *Toronto Star*, 29 September 2016. https://www.thestar.com/amp/news/canada/2017/09/29/ottawas-netflix-deal-angers-culture-conscious-quebec.html.

Heffernan, Virginia. *Magic and Loss: The Internet as Art.* New York: Simon & Schuster, 2016.

Hendy, David. *Public Service Broadcasting.* Basingstoke: Palgrave Macmillan, 2013.

Hindman, Matthew. *The Internet Trap: How the Digital Economy Builds Monopolies and Undermines Democracy.* Princeton, NJ: Princeton University Press, 2018.

Holm, Knut Erik. "The Debate about Public Broadcasting in the Modern World: A Case Study of NRK." Research paper. Oxford: Reuters Institute for the Study of Journalism, University of Oxford, 2014.

Holtz-Bacha, Christina. "The Role of Public Service Media in Nation-Building." In *Public Service Media in Europe: A Comparative Approach*, edited by Karen Arriaza Ibarra, Eva Nowak, and Raymond Kuhn, 27–40. London: Routledge, 2015.

Houpt, Simon. "Ex-journalist Tom Clark, Actor Colin Feore among CBC Board Advisory Team Unveiled by Joly." *Globe and Mail*, 20 June 2017. https://www.theglobeandmail.com/arts/television/heritage-minister-melanie-joly-unveils-cbc-board-advisory-team/article35404032/.

Houpt, Simon, and Susan Krashinsky. "CBC Ad-Free Proposal Shakes Up Heritage Discussion." *Globe and Mail*, 2 December 2016. http://www.theglobeandmail.com/report-on-business/cbcs-ad-free-proposal-shakes-up-heritage-discussion/article33186433/.

Jacoby, Susan. *Why Baseball Matters.* New Haven, CT: Yale University Press, 2018.

Jenkins, Henry, Sam Ford, and Joshua Green. *Spreadable Media: Creating Value and Meaning in a Networked Culture.* New York: New York University Press, 2013.

Jenkins, Henry, Mizuko Ito, and danah boyd. *Participatory Culture in a Networked Era: A Conversation on Youth, Learning, Commerce, and Politics* Cambridge: Polity, 2016.

Jokinen, Tom. "Is the Sun Rising or Setting on the CBC?" *Walrus*, September 2017. https://thewalrus.ca/is-the-sun-rising-or-setting-on-the-cbc/.

Kafka, Peter. "The NFL Can't Blame Trump Any More: It Is Facing a 'Structural Decline in Viewership.'" *recode*, 29 January 2018.

Katz, Elihu. "And Deliver Us from Segmentation." *Annals of the American Academy of Political and Social Science* 546, no. 1 (July 1996): 22–33.

Kessler, Carson. "MLB Attendance Drops to Lowest Average in 15 Years." *Fortune*, 15 June 2018.

Kiefl, Barry. "CBC Must Abandon Ads and Find News Sources of Funding." *Huffington Post*, 10 February 2016. https://www.huffingtonpost.ca/barry-kiefl/cbc-advertising-revenues_b_9201924.html.

– "CBC's Future Should Look Like Its Radio Past." *Globe and Mail*, 6 January 2014.

– "Cutting Government Appointees Could Stall CBC's Downwards Spiral." *Huffington Post*, 19 September 2016. http://www.huffingtonpost.ca/barry-kiefl/cbc-problems_b_12049472.html.

– "Has Canada Lost Faith in the CBC?" *Huffington Post*, 28 July 2014. http://www.huffingtonpost.ca/barry-kiefl/cbc-public-opinion_b_5404479.html.

– "On Treacherous Ground." *Literary Review of Canada* 22, no. 5 (June 2014). https://reviewcanada.ca/magazine/2014/06/on-treacherous-ground/.

Krashinsky Robertson, Susan, and Laura Stone. "Liberal Deal Lets Netflix Play by Its Own Rules in Canadian Broadcasting." *Globe and Mail*, 30 September 2017. https://beta.theglobeandmail.com/report-on-business/liberal-deal-puts-netflix-on-its-own-plane-in-canadian-broadcasting/article36449302/.

Kreiss, Daniel. "The Media Are about Identity, Not Information." In *Trump and the Media*, edited by Pablo Boczkowski and Zizi Papacharissi, 93–99. Cambridge, MA: MIT Press, 2018.

Lawton, Valerie. "APEC Coverage Was Fair, CBC Says: Broadcaster Strongly Rejects Accusations from PM's Office." *Toronto Star*, 7 November 1998.

Levitsky, Steven, and Daniel Ziblatt. *How Democracies Die*. New York: Crown, 2018.

Losse, Kate. "The Real 'News Curators' at Facebook Are the Engineers Who Write Its Algorithms." *Real Future*, 5 October 2016. Accessed 10 May 2018. https://splinternews.com/the-real-news-curators-at-facebook-are-the-engineers-wh-1793856718.

Lotz, Amanda D. *We Now Disrupt This Broadcast: How Cable Transformed Television and the Internet Revolutionized It All*. Cambridge: MA: MIT Press, 2018.

Lozo, Dave. "The NHL Has a Problem: All Its Fans Are Bloody Old." *VICE Sports*, 8 June 2017.

Martin, Don. "Minister Prefers Ad-Less CBC." *National Post*, 17 March 2009.

McChesney, Robert. *Digital Disconnect: How Capitalism Is Turning the Internet against Democracy*. New York: New Press, 2013.

Meisel, John. "Escaping Extinction: Cultural Defence of an Undefended Border." In *Southern Exposure: Canadian Perspectives on the United States*, edited by David Flaherty and William McKercher, 152. Toronto: McGraw Hill Ryerson, 1982.

Millette, Josianne, Mélanie Millette, and Serge Proulx. *Attachement des communautés culturelles aux médias: Le cas des communautés haïtienne, italienne et maghrébine de la région de Montréal*. Les Cahiers-médias no. 19. Sainte-Foy, QC: Centre d'études sur les médias, Université Laval, 2010.

Mounk, Yascha. *The People vs. Democracy: Why Our Freedom Is in Danger and How to Save It*. Boston, MA: Harvard University Press, 2018.

Nash, Knowlton. *The Microphone Wars: A History of Triumph and Betrayal at the CBC*. Toronto: McClelland & Stewart 1994.

Ngao, Erica. "The Podcast Evolution." *Ryerson Review of Journalism* 43, no. 1 (Spring 2017): 30–37.

Nordicity. *Analysis of Government Support for Public Broadcasting*. Prepared for CBC/Radio-Canada. 11 April 2016.

Pallotta, Frank. "NFL Ratings Rebound after Two Seasons of Declining Viewership," *CNN Business*, 3 January 2019. https://www.cnn.com/2019/01/03/media/nfl-ratings-2018-season/index.html.

Pariser, Eli. *The Filter Bubble: What the Internet Is Hiding from You*. New York: Penguin, 2011.

Patterson, Thomas. *Out of Order*. New York: Knopf, 1993.

Prior, Markus. *Post-broadcast Democracy*. New York: Cambridge University Press, 2007.

Public Policy Forum. *The Shattered Mirror: News, Democracy and Trust in the Digital Age*. Ottawa: Public Policy Forum, January 2017. https://shatteredmirror.ca/.

Putnam, Robert. *Bowling Alone: The Collapse and Revival of American Community*. New York: Simon & Schuster 2000.

Raboy, Marc. *Missed Opportunities: The Story of Canada's Broadcasting Policy*. Montreal and Kingston: McGill-Queen's University Press, 1990.

Ramsey, Phil. "Public Service Media Funding in Ireland Faces Continuing Challenges." *Media Policy Project* (blog). London School of Economics and Political Science, 26 September 2017. https://blogs.lse.ac.uk/media policyproject/2017/09/06/public-service-media-funding-in-ireland-faces-continuing-challenges/.

Rasporich, Beverly J. *Made-in-Canada Humour: Literary, Folk and Popular Culture*. Amsterdam: John Benjamins Publishing, 2015.

Romano, Andrew. "The Way They Hook Us—for 13 Hours Straight." *Daily Beast*, May 2013. Accessed 22 April 2018. http://www.andrewromano.net/40/the-way-they-hook-us-for-13-hours-straight.

Rowland, Wade. *Canada Lives Here: The Case for Public Broadcasting*. Westmount, QC: Linda Leith Publishing, 2015.

– *Saving the CBC: Balancing Profit and Public Service*. Westmount, QC: Linda Leith Publishing, 2013.

RTÉ. "RTÉ Broadcasts 15 of Top TV Programmes of 2015." News release. 7 October 2017.

Russell, Nicholas. *Morals and the Media: Ethics in Canadian Journalism.* Vancouver: UBC Press, 2006.

Rutherford, Paul. *When Television Was Young: Primetime Canada, 1952–1967.* Toronto: University of Toronto Press, 1990.

Sandomir, Richard. "With Audiences Shrinking, NBC Looks Cautiously to Olympics in Asia." *New York Times,* 22 August 2016. http://www.nytimes.com/2016/08/23/sports/with-audience-shrinking-nbc-looks-cautiously-to-olympics-in-asia.html?smprod=nytcore-iphone&smid=nytcore-iphone-s.

Saulnier, Alain. *Losing Our Voice: Radio-Canada under Siege.* Toronto: Dundurn, 2015.

Sauvageau, Florian. Keynote address presented at the "How Canadian Communicate Politically: The Next Generation" conference, Banff, AB, October 2009.

– "Millennium Blues: The 1997 Southam Lecture." *Canadian Journal of Communication* 23, no. 2 (February 1998). https://doi.org/10.22230/cjc.1998v23n2a1029.

Schneller, Johanna. "A Dip into the Wide (Apple TV +) and Narrow (CBC's Gem) of Streaming Services." *Globe and Mail,* 1 April 2019. Accessed 3 April 2019. https://www.theglobeandmail.com/arts/article-cbcs-gem-originals-are-pure-kind-canadiana/.

Shoalts, David. *Hockey Fight in Canada.* Madeira Park, BC: Douglas & McIntyre, 2018.

– "Top Hockey Producer's Firing before NHL Playoffs Hints at Rogers Turmoil." *Globe and Mail,* 7 April 2016. https://www.theglobeandmail.com/sports/hockey/top-hockey-producers-firing-before-nhl-playoffs-hints-at-rogers-turmoil/article29548827/.

Smith, Michael, and Rahul Telang. *Streaming, Sharing, Stealing: Big Data and the Future of Entertainment.* Cambridge, MA: MIT Press, 2016.

Speirs, Rosemary. "PM Gets a Rough Ride—and Shows a Weak Spot." *Toronto Star,* 12 December 1996.

Sports Media Watch. "National Hockey League Stanley Cup Games Average TV Ratings in the U.S. from 2006 to 2019." Accessed 16 August 2018. https://www.statista.com/statistics/305818/average-tv-ratings-nhl-stanley-cup-games/.

Statista.com. "Average TV Viewership of NBA Finals in the United States from 2002 to 2018." https://www.statista.com/statistics/240381/nba-finals-tv-ratings-in-the-united-states/.

– "Leading Social Networks Used Weekly for News in Canada as of February 2018." Accessed 27 March 2019. https://www.statista.com/statistics/563514/social-networks-used-for-news-canada/.

Stinson, Scott. "CBC's Strange Partnership with Rogers to Show NHL Games Remains at Odds with Identity as Public Broadcaster." *National Post*, 25 May 2017. https://nationalpost.com/sports/hockey/nhl/cbcs-strange-partnership-with-rogers-to-show-nhl-games-remains-at-odds-with-identity-as-public-broadcaster.

Stone, Linda. "Just Breathe: Building the Case for Email Apnea." *Huffington Post*, 2 August 2008. https://www.huffpost.com/entry/just-breathe-building-the_n_85651.

Strachin, Jamie. "Cable TV Cord Cutters Threaten Sports Business." *cbc.ca*, 13 May 2017. http://www.cbc.ca/sports/cable-sports-tv-cord-cutting-1.4112895.

Stursberg, Richard. "The CBC Is Dying: Here's How to Save It." *Toronto Star*, 19 November 2015. http://www.thestar.com/opinion/commentary/2015/11/19/the-cbc-is-dying-heres-how-to-save-it.html.

– *The Tower of Babble: Sins, Secrets and Successes inside the CBC*. Vancouver: Douglas & McIntyre, 2012.

Stursberg, Richard, with Stephen Armstrong. *The Tangled Garden: A Canadian Cultural Manifesto for the Digital Age*. Toronto: James Lorimer, 2019.

Sunstein, Cass. *#Republic: Divided Democracy in the Age of Social Media*. Princeton, NJ: Princeton University Press, 2017.

Sweazey, Connor. "Broadcasting Canada's War: How the Canadian Broadcasting Corporation Reported the Second World War." MA thesis, University of Calgary, 2017.

Syvertsen, Trine, Gunn Enli, Ole Mjos, and Hallvard Moe. *The Media Welfare State: Nordic Media in the Digital Era*. Ann Arbor: University of Michigan Press, 2014.

Taber, Jane. "CBC Clears Pollster, Criticizes 'Paranoia-Tinged' Tories." *Globe and Mail*, 19 May 2010.

Tait, Catherine. "CBC/Radio-Canada: At the Heart of Canadians' Lives." Speech presented to the Montreal Chamber of Commerce, 3 May 2019. https://cbc.radio-canada.ca/en/media-centre/2019/speaking-notes-catherine-tait-chamber-commerce-mtl.

Taras, David. *Digital Mosaic: Media, Power, and Identity in Canada*. Toronto: University of Toronto Press, 2015.

– "The Mass Media and Political Crisis: Reporting Canada's Constitutional Struggles." *Canadian Journal of Communication* 18, no. 2 (Spring 1993): 131–48.

– *Power and Betrayal in the Canadian Media*. Peterborough: Broadview Press, 2001.

– "The Struggle over *The Valour and the Horror*: Media Power and the Portrayal of War." *Canadian Journal of Political Science* 28, no. 4 (December 1995): 725–48. https://doi.org/10.1017/S0008423900019363.

Taras, David, and Christopher Waddell. *How Canadians Communicate V: Sports*. Edmonton: Athabasca University Press, 2016.

Taylor, Gregory. "Canadian Journalism Is in Crisis: CBC News Can Save It." *Huffington Post*, 28 June 2017. https://www.huffingtonpost.ca/ gregory-taylor/canadian-journalism-is-in-crisis-cbc-news-can-save-it_a_23004530/.

Taylor, Kate. "Ad-Free CBC Could Serve as Rallying Point for Canadian Creativity." *Globe and Mail*, 3 December 2016. https://beta.theglobeandmail. com/arts/television/ad-free-cbc-could-act-as-a-rallying-point-for-canadian-creativity/article33138985/?ref=http://www.theglobeandmail.com.

Thompson, Derek. *Hit Makers: The Science of Popularity in an Age of Distraction*. New York: Penguin, 2017.

– "Why NFL Ratings Are Plummeting: A Two-Part Theory." *Atlantic*, 1 February 2018.

Toronto Star. "The Prime Minister Is Lying." Editorial. 12 December 1996.

Travis, Clay. "ESPN Loses 500,000 Subscribers in April." Outkick the Coverage, 1 May 2018. https://www.outkickthecoverage.com/ espn-loses-500000-subscribers-april/.

Turkle, Sherry. *Reclaiming Conversation*. New York: Penguin, 2015.

van Dijck, José. *The Culture of Creativity: A Critical History of Social Media*. New York: Oxford University Press, 2013.

Vipond, Mary. *The Mass Media in Canada*. Toronto: James Lorimer, 2000.

Waddell, Christopher. "Government Funding for Journalism: To What End?" *The Conversation*, 21 March 2019. https://theconversation.com/ government-funding-for-journalism-to-what-end-113978.

– "The Hall of Mirrors." In *How Canadians Communicate V: Sports*, edited by David Taras and Christopher Waddell, 41–54. Edmonton: Athabasca University Press, 2016. http://www.aupress.ca/books/120244/ebook/99Z_ Taras_Waddell_2016-How_Canadians_Communicate_V_Sports.pdf.

Watson, Patrick. *This Hour Has Seven Decades*. Toronto: McArthur and Company, 2004.

Watters, Hayden. "Vote Compass Has Been Used 365K Times: What Has Been Learned?" cbc.ca, 6 June 2018. https://www.cbc.ca/news/canada/toronto/ live-blog-vote-compass-1.4694038.

Webster, James. *The Marketplace of Attention: How Audiences Take Shape in a Digital Age*. Cambridge, MA: MIT Press, 2014.

Weiner, Jonah. "The Great Race to Rule Streaming TV." *New York Times Magazine*, 10 July 2019.

Wikipedia. "List of Most Watched Television Broadcasts in Canada." Accessed 15 August 2018. https://en.wikipedia.org/wiki/ List_of_most_watched_television_broadcasts_in_Canada.

Winn, Ross. "2019 Podcast Stats & Facts." 6 March 2019. https://www. podcastinsights.com/podcast-statistics/.

Wolff, Michael. *Television Is the New Television*. New York: Portfolio, 2015.

Wu, Tim. *The Attention Merchants*. New York: Knopf, 2016.

– *The Master Switch: The Rise and Fall of Information Empires*. New York: Knopf, 2010.

Yao, Richard. "The Unraveling of Live Sports TV." IPG Media Lab, 22 February 2018.

Zerbesias, Antonia. "APEC Squabble Could Cost CBC." *Toronto Star*, 20 October 1998.

Index